SYMMETRY–ADAPTED BASIS SETS

BASIS SETS

Automatic Generation for Problems in Chemistry and Physics

SYMMETRY-ADAPTED BASIS SETS

Automatic Generation for Problems in Chemistry and Physics

John Scales Avery
Sten Rettrup
University of Copenhagen, Denmark

James Emil Avery
NZIAS, Massey University, New Zealand

NEW JERSEY · LONDON · SINGAPORE · BEIJING · SHANGHAI · HONG KONG · TAIPEI · CHENNAI

Published by

World Scientific Publishing Co. Pte. Ltd.

5 Toh Tuck Link, Singapore 596224

USA office: 27 Warren Street, Suite 401-402, Hackensack, NJ 07601

UK office: 57 Shelton Street, Covent Garden, London WC2H 9HE

British Library Cataloguing-in-Publication Data
A catalogue record for this book is available from the British Library.

SYMMETRY-ADAPTED BASIS SETS
Automatic Generation for Problems in Chemistry and Physics

Copyright © 2012 by World Scientific Publishing Co. Pte. Ltd.

ISBN-13 978-981-4350-46-4
ISBN-10 981-4350-46-X

Printed in Singapore by World Scientific Printers.

Contents

8. SYMMETRY-ADAPTED SOLUTIONS BY ITERATION   121

    8.1 Conservation of symmetry under Fourier transformation . . 121
    8.2 The operator $-\Delta + p_\kappa^2$ and its Green's function 122
    8.3 Conservation of symmetry under iteration of the
         Schrödinger equation 123
    8.4 Evaluation of the integrals 125
    8.5 Generation of symmetry-adapted basis functions by iteration 127
    8.6 A simple example 128
    8.7 An alternative expansion of the Green's function that
         applies to the Hamiltonian formulation of physics 129

Appendix A REPRESENTATION THEORY OF FINITE GROUPS 131

    A.1 Basic definitions 131
    A.2 Representations of geometrical symmetry groups 132
    A.3 Similarity transformations 133
    A.4 Characters and reducibility 134
    A.5 The great orthogonality theorem 136
    A.6 Classes 141
    A.7 Projection operators 142
    A.8 The regular representation 144
    A.9 Classification of basis functions 146

Appendix B STURMIAN BASIS SETS   149

    B.1 One-electron Coulomb Sturmians 149
    B.2 Löwdin-orthogonalized Coulomb Sturmians 151
    B.3 The Fock projection 153
    B.4 Generalized Sturmians and many-particle problems 154
    B.5 Use of generalized Sturmian basis sets to solve the many-
         particle Schrödinger equation 155
    B.6 Momentum-space orthonormality relations for Sturmian
         basis sets 156
    B.7 Sturmian expansions of d-dimensional plane waves 158
    B.8 An alternative expansion of a d-dimensional plane wave . . 159

Appendix C ANGULAR AND HYPERANGULAR INTEGRATION 161

    C.1 Monomials, homogeneous polynomials, and harmonic
         polynomials 161
    C.2 The canonical decomposition of a homogeneous polynomial 162

Preface

In a number of previous papers and a book [Avery and Avery, 2006], two of us introduced the Generalized Sturmian Method applied to atoms. One of the interesting features of this method is that it leads to a secular equation of the form

$$\sum_{\nu} \left[\delta_{\nu',\nu} Z \mathcal{R}_\nu + T'_{\nu',\nu} - p_\kappa \delta_{\nu',\nu} \right] C_{\nu,\kappa} = 0 \tag{0.1}$$

where Z is the nuclear charge and where p_κ is not the energy but is a scaling factor related to the energy of an atomic state by

$$E_\kappa = -\frac{p_\kappa^2}{2} \tag{0.2}$$

In equation (0.1),

$$\mathcal{R}_\nu \equiv \sqrt{\frac{1}{n^2} + \frac{1}{n'^2} + \frac{1}{n''^2} + \ldots} \tag{0.3}$$

ν being a collection of indices that label a configuration (a Slater determinant) and $n, n', n'', \ldots$ being the principal quantum numbers of the atomic spin-orbitals that appear in the configuration. The matrix $T'_{\nu',\nu}$ represents the effects of interelectron repulsion. If interelectron repulsion is entirely neglected, the energy reduces to

$$E_0 = -\frac{1}{2}(Z\mathcal{R}_\nu)^2 = -\frac{1}{2}Z^2 \left(\frac{1}{n^2} + \frac{1}{n'^2} + \frac{1}{n''^2} + \ldots \right) \tag{0.4}$$

that is to say, it reduces to the energy of a set of completely independent electrons moving in the attractive Coulomb potential of the nucleus. We introduced a large-Z approximation, in which the basis set is restricted to configurations all of which correspond to the same value of R_ν, and we found that in this approximation, the energy of an atomic state is given by

$$E_\kappa \approx -\frac{1}{2}(Z\mathcal{R}_\nu - |\lambda_\kappa|)^2 \tag{0.5}$$

where λ_κ is a root of the energy-independent interelectron repulsion matrix $T'_{\nu',\nu}$. We found that for an isoelectronic series corresponding to a fixed number of electrons N, the large-Z approximation approached the experimental data more and more closely with increasing Z, provided that a simple correction was made for relativistic effects. We also found that the eigenfunctions of the simplified secular equation

$$\sum_{\nu} \left[T'_{\nu',\nu} - \lambda_\kappa \delta_{\nu',\nu} \right] C_{\nu,\kappa} = 0 \qquad (0.6)$$

were Russell-Saunders states, i.e. simultaneous eigenfunctions of the total angular momentum operator L^2, its z-component L_z, the total spin operator S^2, and its z-component S_z. These Russell-Saunders states could be used as symmetry-adapted basis functions for more refined calculations, where the large-Z approximation was abandoned. This automatic generation of symmetry-adapted basis functions proved to be a great convenience.

In the present book, we are joined by Lektor Sten Rettrup, who has taught Group Theory and Quantum Chemistry for many years at the University of Copenhagen. The book results from a wish to go more deeply into the mathematical reasons behind the automatic generation of symmetry-adapted basis sets, which we observed when applying the Generalized Sturmian Method to atoms in the large-Z approximation. We believe that the same principles can be used in many problems encountered in physics and chemistry, where the size of the basis sets required for accurate solutions makes it desirable to use symmetry-adapted basis functions.

The first chapter discusses the general mathematical principles that underlie the automatic generation of symmetry-adapted basis functions, and subsequent chapters aim at developing and illustrating these basis principles. In Chapter 2, the principles are applied to atomic Hartree-Fock calculations followed by Configuration Interaction computations, while in Chapter 3, this is extended to molecules. As is well known, the number of configurations needed for accurate results in CI calculations is often extremely large - often in the millions. Therefore reducing this number by means of symmetry-adapted basis functions is highly desirable. We hope that Chapters 2 and 3 indicate paths one might follow to do this.

Chapter 4 reviews and extends our studies of the Generalized Sturmian method applied to atoms. Among the new topics not discussed previously, is the calculation of the effect on atoms when they are surrounded by symmetrical arrays of point charges. This effect is calculated using the Generalized Sturmian Method.

Chapter 5 sketches the extension of the Generalized Sturmian Method to molecules. Coulomb Sturmians have long been used as basis functions for solution for a single electron moving in the attractive field of two or more nuclei. Among the important pioneers in this field are C.E. Wulfman, V. Aquilanti and T. Koga. In Chapter 5, we try to show how Sturmian basis functions can be used in N-electron molecules. To do this, we need to evaluate many-center interelectron repulsion integrals involving Coulomb Sturmians, which are exponential-type orbitals. We propose a new method for evaluating these integrals making use of the Fock projection, which maps Coulomb Sturmians onto sets of 4-dimensional hyperspherical harmonics. The properties of hyperspherical harmonics are then used to evaluate the needed integrals.

In Chapters 6 and 7, we attempt to show how the basic principles for automatic generation of symmetry-adapted basis functions may be applied to other areas of physics. We hope that readers will find ways of applying these general principles to their own special areas of research. Chapter 8 discusses how symmetry-adapted basis functions can be generated by iteration.

Atomic units are used throughout this book: All lengths are in Bohrs and all energies are in Hartrees.

Programs for carrying out calculations of the type discussed in this book can be found at our website [Avery and Avery, 2006a]

`http://sturmian.kvante.org/symmetry`

These programs may be freely downloaded and modified for general use.

John Scales Avery, Sten Rettrup and James Emil Avery
Copenhagen, 2011

Chapter 1

GENERAL CONSIDERATIONS

1.1 The need for symmetry-adapted basis functions

Many problems in physics, theoretical chemistry, theoretical astronomy and engineering essentially reduce to solving partial differential equations that give rise to eigenvalue problems of the form

$$T \left| \Psi_\kappa \right\rangle = \lambda_\kappa \left| \Psi_\kappa \right\rangle \tag{1.1}$$

subject to a set of boundary conditions. Here, T is a differential operator, λ_κ denotes the κth eigenvalue, and $\left| \Psi_\kappa \right\rangle$ is the κth eigenfunction. For example, the differential operator T in equation (1.1) might be the Hamiltonian of a many-particle system, $\left| \Psi_\kappa \right\rangle$ representing the wave function and λ_κ being the energy.

When the eigenvalue problem is difficult or impossible to solve exactly (as it is in the example just mentioned), one usually seeks approximate solutions in a finite dimensional subspace to the full solution space. Given a basis $\{ \left| \Phi_1 \right\rangle, \left| \Phi_2 \right\rangle, \dots, \left| \Phi_M \right\rangle \}$ for the approximate space, this reduces (1.1) to the problem of diagonalizing the corresponding matrix representation:

$$T_{\nu', \nu} \equiv \left\langle \Phi_{\nu'} | T | \Phi_\nu \right\rangle \tag{1.2}$$

If the basis set is orthonormal, one solves the secular equations

$$\sum_{\nu=1}^{M} \left[T_{\nu', \nu} - \lambda_\kappa \delta_{\nu', \nu} \right] C_{\nu\kappa} = 0 \tag{1.3}$$

to obtain the solutions $\left| \Psi_\kappa \right\rangle$ as superpositions:

$$\left| \Psi_\kappa \right\rangle = \sum_{\nu=1}^{M} \left| \Phi_\nu \right\rangle C_{\nu\kappa} \tag{1.4}$$

However, while the overall procedure outlined above is exceedingly simple, in many cases the number M of basis functions needed to obtain accurate solutions is very large, causing both construction and diagonalization of $T_{\nu'\nu}$ to be arduous tasks. This difficulty has led to the invention of a great many approximation methods that reduce the dimensionality, usually highly tailored to the class of physical problems under study. *Symmetry methods*, the topic of this book, are both generic and problem specific in this sense: The general framework and methods are agnostic towards the underlying physical problem, but the specific symmetry groups and their actions on the solution space are particular to the problem at hand. The methods do not stand in the way of other methods of reducing complexity, which can be applied before or after reduction through symmetry.

The use of symmetry in chemistry and physics is a large and well-studied field. There are many excellent textbooks that teach well-established methods for reducing and analyzing physical problems. However, most methods require a great deal of human work and creativity. The methods on which we will focus in this book, on the other hand, have automation as a primary goal. We strive to automate not only the reduction in complexity of the original physical problems, but also subsequent classification of the resulting solutions according to their symmetries.

1.2 Fundamental concepts

Consider a differential operator $T : V \to V$ on a Sobolev space V. A *symmetry operation* of T is a transformation of the underlying space V that leaves the eigensolutions to T invariant. That is, it is an automorphism[1] G of V such that $T|\Psi\rangle = \lambda|\Psi\rangle$ if and only if $TG|\Psi\rangle = \lambda G|\Psi\rangle$. It is easy to show, using the fact that the eigenvectors of T form a basis for V, that this property is fulfilled exactly when the automorphism G commutes with T, i.e. if and only if

$$GT = TG \qquad (1.5)$$

Symmetry operations have an inherent group structure and arise naturally as actions on V of an underlying *symmetry group* $\mathcal{G}$ that is implicitly defined by the physical problem. The group action constitutes a *group representation* of $\mathcal{G}$ on the space V, and because of this, representation theory is a central tool of symmetry analysis. We give an introduction to the relevant

---

[1] An invertible linear function from V into itself.

aspects of group representation theory in Appendix A. Readers who are not, or only superficially, familiar with representation theory are encouraged to study this appendix before proceeding to Chapter 2. While we try to keep the main part of the book free of the intricacies and technical jargon of representation theory to the extent that it does not impede understanding, all of our methods do depend heavily on its central results.

For many physical problems, the underlying symmetry group is either *finite* or *compact*. When this is the case, any representation can be broken down into a direct sum of finite dimensional irreducible representations, which are representations acting on smallest invariant subspaces. A *basis* for an irreducible representation is a basis for the subspace on which it acts. In this form, the representation is maximally block diagonal, i.e. the action of every group element is block diagonal with respect to these smallest invariant subspaces. Fortunately, because of the great orthogonality theorem (and its generalization to compact topological groups), the irreducible representations also induce block diagonality of the operator. This defines a very useful decomposition of the eigenproblem into completely independent subproblems. In addition, each eigenvector to T transforms according to a unique symmetry, which gives us a means to classify solutions to the eigenproblem.

Decomposing the symmetry group action into irreducible representations and finding the associated smallest invariant spaces, however, is generally not an easy task, and it requires a high level of ingenuity to be applied for each class of problems under study. It is frequently much easier to find invariant subspaces that are small, but not minimal. Oftentimes, a chosen basis set for the solution space has structure that allows such subspaces to be trivially and automatically derived directly from the basis set. We will show how an initial decomposition into non-minimal spaces can be used to automatically compute a full decomposition, resulting in a transformation of the original basis set into symmetry adapted basis functions, with respect to which the eigenproblem breaks down into independent subproblems. What is more, we can pick a particular symmetry and choose to compute only those solutions that correspond to it.

1.3 Definition of invariant blocks

Let us now consider a finite dimensional Hilbert space W with the basis $\{|\Phi_1\rangle, \ldots, |\Phi_M\rangle\}$, and let $T : W \to W$ be a Hermitian operator on W with the symmetry group $\mathcal{G}$. We can re-order and partition the basis set so as to produce a number of disjoint subsets, each of which is closed under the operations of $\mathcal{G}$. Basis sets are often defined in such a way that we can do this directly. However, even if we are given no apparent structure, we can always produce such a partition of the basis by first acting on $|\Phi_1\rangle$ with all the elements of $\mathcal{G}$ and successively acting on the resulting sets until closure is reached. The remaining basis functions will span the orthogonal complement to the invariant space, and the process can be repeated until no more basis functions remain. This yields a partition of the original basis into *invariant subsets* $\{|\Phi_\nu^1\rangle\}, \ldots, \{|\Phi_\nu^p\rangle\}$, such that their linear spans $W_1, \ldots, W_p$ each are closed under all the actions of $\mathcal{G}$. It is, of course, very laborious to construct the W_k's by acting on the basis functions with the elements of $\mathcal{G}$, but fortunately this is rarely necessary. In almost all cases, the original basis partitions naturally into invariant subsets, as will be illustrated in Chapters 2-7.

Let us call the length of the kth subset m_k, and the $\mathcal{G}$-invariant subspace that it spans W_k. We then have

$$m_1 + \cdots + m_p = M \tag{1.6}$$

and

$$W = W_1 \oplus \cdots \oplus W_p \tag{1.7}$$

We will call the small block of $T_{\nu',\nu}$ based on this subset an *invariant block*, $T_{\nu',\nu}^k \equiv \langle \Phi_{\nu'}^k | T | \Phi_\nu^k \rangle$. The same name is given to the restrictions $T^k : W_k \to W_k$ of the operator T. These are defined in terms of the projection operators onto the spaces W_k as

$$T^k \equiv P_{W_k} T P_{W_k} \tag{1.8}$$

allowing us to speak of invariant blocks in a basis independent way.

The original basis set $\{|\Phi_1\rangle, |\Phi_2\rangle, \ldots, |\Phi_M\rangle\}$ spanning the Hilbert space W is assumed to be closed under the operations of $\mathcal{G}$, and thus according to Weyl's Theorem, the representation of $\mathcal{G}$ on it is fully reducible [Weyl, 1950; Weyl, 1939]. Because of this, there exists a basis set for W consisting of symmetry adapted functions; see Appendix A. The same is true for the invariant subspaces W_k into which the original Hilbert space W is decomposed, because each subspace is constructed in such a way that it is closed under the operations of $\mathcal{G}$.

1.4 Diagonalization of the invariant blocks

While the operator T itself is not at all block diagonal with respect to the invariant subsets of the basis, the small invariant blocks $T^k : W_k \to W_k$ of T do have one important property: their eigenfunctions are symmetry-adapted and can be used to break down the full eigenproblem into smaller subproblems.

We would like to show that the eigenfunctions of the small invariant blocks T^k are symmetry-adapted basis functions, the use of which will facilitate the diagonalization of the full operator T.

To show that this is true, let us consider an orthonormal set of symmetry-adapted basis functions $|\eta_j^{\alpha,n}\rangle$ spanning the invariant subspace W_k. By the same argument that a symmetry adapted basis exists for W, such a basis set exists for each each of the subspaces W_k: The subspaces are closed under the actions of $\mathcal{G}$, and hence by Weyl's Theorem [Weyl, 1950], the representations of $\mathcal{G}$ on them are fully reducible. Thus we can write a symmetry adapted matrix representation of the invariant block operators T^k:

$$
\begin{aligned}
T^k_{(\alpha',n',j),(\alpha,n,j)} &\equiv \left\langle \eta_{j'}^{\alpha',n'} \middle| T^k \middle| \eta_j^{\alpha,n} \right\rangle \\
&= \left\langle \eta_{j'}^{\alpha',n'} \middle| P_{W_k} T P_{W_k} \middle| \eta_j^{\alpha,n} \right\rangle = \left\langle \eta_{j'}^{\alpha',n'} \middle| T \middle| \eta_j^{\alpha,n} \right\rangle
\end{aligned}
\tag{1.9}
$$

The index α labels the inequivalent irreducible representations, while n labels the basis functions within an irreducible representation. The final index j labels different basis functions of the same symmetry.

We prove in Appendix A that symmetry-adapted basis functions have the special property that

$$
\left\langle \eta_{j'}^{\alpha',n'} \middle| T \middle| \eta_j^{\alpha,n} \right\rangle = \delta_{n',n} \delta_{\alpha',\alpha} \left\langle \eta_{j'}^{\alpha',n'} \middle| T \middle| \eta_j^{\alpha,n} \right\rangle
\tag{1.10}
$$

In other words, a matrix element of T based on the symmetry-adapted functions $|\eta_j^{\alpha,n}\rangle$ is zero unless $\alpha' = \alpha$ and $n' = n$. However, by Equation (1.9), also the matrix $T^k_{(\alpha',n',j'),(\alpha,n,j)}$ is block diagonal with respect to α and n. We will now use this fact to show that any basis of eigenfunctions to T^k is automatically symmetry adapted.

Let $\{|\omega_\kappa\rangle\}$ be some orthonormal basis for W_k consisting of eigenfunctions to the invariant block T^k, and let $U_{(\alpha,n,j),\kappa}$ be the unitary transformation matrix that brings the η-basis into the ω-basis:

$$|\omega_\kappa\rangle = \sum_{(\alpha,n,j)} |\eta_j^{\alpha,n}\rangle U_{(\alpha,n,j),\kappa} \tag{1.11}$$

By definition, $U_{(\alpha,n,j),\kappa}$ diagonalizes $T^k_{(\alpha',n',j'),(\alpha,n,j)}$, i.e.

$$
\begin{aligned}
(U^\dagger T^k U)_{\kappa'\kappa} &\equiv \sum_{(\alpha',n',j'),(\alpha,n,j)} U^\dagger_{\kappa',(\alpha',n',j')} T^k_{(\alpha',n',j'),(\alpha,n,j)} U^\dagger_{\kappa,(\alpha,n,j)} \\
&= \langle \omega_{\kappa'} | T^k | \omega_\kappa \rangle \\
&= \delta_{\kappa'\kappa} \lambda_\kappa
\end{aligned}
\tag{1.12}
$$

But diagonalizing a block diagonal matrix is the same as diagonalizing each block separately, so the transformation cannot mix elements from the different (α, n)-blocks of $T^k_{(\alpha',n',j'),(\alpha,n,j)}$. We consequently obtain

$$|\omega_\kappa\rangle = \sum_j |\eta_j^{\alpha,n}\rangle U_{(\alpha,n,j),\kappa} \tag{1.13}$$

That is, the eigenfunction basis $\{|\omega_\kappa\rangle\}$ is also symmetry adapted with respect to α and n, regardless of how the invariant block operator T^k was diagonalized. In the case where there are no accidental degeneracies, the functions $\{|\omega_\kappa\rangle\}$ are basis functions for irreducible representations of the group $\mathcal{G}$. If there are accidental degeneracies, the representations may be reducible.

When degeneracies occur in the diagonalization of the small invariant blocks, one can introduce an extremely small perturbation T' to the operator T^k that very slightly removes the degeneracies. Then each of the eigenfunctions $|\omega_\kappa\rangle$ obtained by diagonalizing an invariant block of $T^k_{\nu',\nu}$ is unique, and with the reduction of symmetry caused by the perturbation, the representations are 1-dimensional. The type of perturbation needed for this purpose will be discussed later in connection with particular examples.

1.5 Transformation of the large matrix to block-diagonal form

Let $S_{\nu,\kappa}$ be the $M \times M$ dimensional similarity transformation matrix bringing the matrix $T_{\nu',\nu}$ from the original representation (1.2) to a representation based on eigenfunctions of the invariant blocks $T^k_{\nu',\nu}$. ($S_{\nu,\kappa}$ is a direct sum of the small $m_k \times m_k$-dimensional transformation matrices $C_{\nu,\kappa}$.) If we let

$$|\omega_\kappa\rangle = \sum_{\nu=1}^M |\Phi_\nu\rangle \langle \Phi_\nu | \omega_\kappa \rangle \equiv \sum_{\nu=1}^M |\Phi_\nu\rangle S_{\nu,\kappa} \tag{1.14}$$

then the transformed matrix

$$\tilde{T}_{\kappa',\kappa} \equiv \langle \omega_{\kappa'}|T|\omega_\kappa \rangle = \sum_{\nu'=1}^{M} \sum_{\nu=1}^{M} S^{\dagger}_{\kappa',\nu'} \langle \Phi_{\nu'}|T|\Phi_\nu \rangle S_{\nu,\kappa} \qquad (1.15)$$

will be block diagonal due to the $|\omega_\kappa\rangle$ being symmetry adapted. Each block of $\tilde{T}_{\kappa',\kappa}$ can be solved for independently, and we obtain, with much less computational effort, the same eigenfunctions and eigenvalues that would be obtained by diagonalizing $T_{\nu',\nu}$. Although the transformation shown in equation (1.15) makes $\tilde{T}_{\kappa',\kappa}$ block-diagonal, an appropriate rearrangement of the rows and columns is needed to make this apparent. However, automatic algorithms for such a rearrangement are simple and rapid, and can be found in our Generalized Sturmian Library [Avery and Avery, 2006a].

1.6 Summary of the method

The method for automatic generation of symmetry-adapted basis functions can be summarized as follows:

(1) Partition the basis set into smaller subsets $\{|\Phi_\nu^1\rangle\}, \ldots, \{|\Phi_\nu^p\rangle\}$, such that the corresponding subspaces $W_1, \ldots, W_p$ each are closed under $\mathcal{G}$. We call the small subsets *invariant subsets*, and denote their lengths by m_k, where $m_1 + m_2 + \cdots = M$.
(2) Construct and diagonalize small $m_k \times m_k$ blocks $T_{\nu'\nu}^k = \langle \Phi_{\nu'}^k|T|\phi_\nu^k \rangle$ based on the invariant subsets. These are called *invariant blocks*, and their eigenvectors are symmetry-adapted with respect to symmetry indices α and n.
(3) If degeneracies occur, we may introduce an extremely small perturbation to T that very slightly removes the degeneracies.

The symmetry adapted basis obtained in this manner defines a transformation of the matrix $T_{\nu',\nu}$ into block-diagonal form, yielding a decomposition of the original eigenvalue problem into smaller subproblems that can each be solved independently.

In Chapters 2-7, we apply the general technique to some particular examples in physics and chemistry. In Chapter 8, we discuss how symmetry-adapted basis functions may be obtained by iteration of integral equations.

Chapter 2

EXAMPLES FROM ATOMIC PHYSICS

2.1 The Hartree-Fock-Roothaan method for calculating atomic orbitals

As an example of the general principles discussed above, let us consider the problem of constructing symmetry-adapted configurations to be used in solving the non-relativistic Schrödinger equation of an N-electron atom.

$$H|\Psi_\kappa\rangle = E_\kappa|\Psi_\kappa\rangle \qquad (2.1)$$

In equation (2.1),

$$H = \sum_{i=1}^{N}\left[-\frac{1}{2}\nabla_i^2 - \frac{Z}{r_i}\right] + \sum_{i>j}^{N}\frac{1}{r_{ij}} \qquad (2.2)$$

is the Hamiltonian of the atom. We can represent the wave function $|\Psi_\kappa\rangle$ as a linear superposition of configurations:

$$|\Psi_\kappa\rangle = \sum_{\nu=1}^{M}|\Phi_\nu\rangle C_{\nu,\kappa} \qquad (2.3)$$

Each configuration is a Slater determinant

$$|\Phi_\nu\rangle = |\chi_\mu\chi_{\mu'}\chi_{\mu''}\cdots| \equiv \frac{1}{\sqrt{N!}}\begin{vmatrix} \chi_\mu(\mathbf{x}_1) & \chi_{\mu'}(\mathbf{x}_1) & \chi_{\mu''}(\mathbf{x}_1) & \cdots \\ \chi_\mu(\mathbf{x}_2) & \chi_{\mu'}(\mathbf{x}_2) & \chi_{\mu''}(\mathbf{x}_2) & \cdots \\ \chi_\mu(\mathbf{x}_3) & \chi_{\mu'}(\mathbf{x}_3) & \chi_{\mu''}(\mathbf{x}_3) & \cdots \\ \vdots & \vdots & \vdots & \end{vmatrix} \qquad (2.4)$$

involving atomic orbitals of the form

$$\chi_\mu(\mathbf{x}_i) \equiv \chi_{n,l,m,m_s}(\mathbf{x}_i) \equiv R_{n,l}(r_i)Y_{l,m}(\theta_i,\phi_i)\begin{cases} \alpha_i & m_s = 1/2 \\ \beta_i & m_s = -1/2 \end{cases} \qquad (2.5)$$

The Self-Consistent-Field (SCF) method for the calculation of atomic orbitals was developed by Douglas R. Hartree and by V. Fock. It aims at producing optimal atomic orbitals when the wave function is approximated by a single Slater determinant. An adaptation of the Hartree-Fock method to matrices and basis sets was later developed by Clemens C.J. Roothaan. Some of the first computer calculations on atoms were carried out by Clemens Roothaan and Enrico Clementi at the University of Chicago. In the Hartree-Fock-Roothaan method, one approximates the atomic orbitals χ_μ by a linear combination of basis functions, u_b.

$$\chi_\mu(\mathbf{x}_i) = \sum_b u_b(\mathbf{x}_i)c_{b,\mu} \qquad (2.6)$$

It is convenient to make the set of basis functions orthonormal before the start of the calculation, so that they satisfy

$$\int d^3x_i \; u_a^*(\mathbf{x}_i)u_b(\mathbf{x}_i) = \delta_{a,b} \qquad (2.7)$$

The coefficients in the superposition (2.6) are found by solving the secular equations

$$\sum_a \left[F_{a,b} - \epsilon_\mu \delta_{a,b} \right] c_{b,\mu} = 0 \qquad (2.8)$$

where the *Fock matrix* $F_{a,b}$ is defined by

$$F_{a,b} = h_{a,b} + \sum_{c,d} P_{c,d}\Gamma_{(ab)cd} \qquad (2.9)$$

In equation (2.9), $h_{a,b}$ is a matrix representing the core Hamiltonian

$$h_{a,b} \equiv \int d^3x_1 \; u_a^*(\mathbf{x}_1) \left[-\frac{1}{2}\nabla_1^2 - \frac{Z}{r_1} \right] u_b(\mathbf{x}_1) \qquad (2.10)$$

while $\Gamma_{(ab)cd}$ is the matrix of Coulomb and exchange integrals

$$\Gamma_{(ab)cd} \equiv \int d^3x_1 \int d^3x_2 u_a^*(\mathbf{x}_1)u_c^*(\mathbf{x}_2)\frac{1}{r_{12}}\left(1 - \mathcal{P}_{12}\right)u_d(\mathbf{x}_2)u_b(\mathbf{x}_1) \quad (2.11)$$

while

$$P_{c,d} \equiv \sum_\mu n_\mu c_{c,\mu}^* c_{d,\mu} \qquad (2.12)$$

is the *density matrix*. In equation (2.12), n_μ is an occupation number, which is 1 or 0, depending on whether the atomic orbital χ_μ is filled or empty:

$$n_\mu \equiv \begin{cases} 1 \text{ filled} \\ 0 \text{ empty} \end{cases} \qquad (2.13)$$

Empty atomic orbitals result from the solution of the secular equations (2.8) because the number of basis functions is larger than the number of electrons. The empty orbitals are called *virtual orbitals*. The atomic orbitals that are calculated by solving the Hartree-Fock-Roothaan equations have the form shown in equation (2.5), the angular and spin parts being fixed, while the radial parts $R_{n,l}(r_i)$ are determined by iterative solution of the equations. One begins by calculating the matrices h_{ab} and $\Gamma_{(ab)cd}$, and by guessing initial values for the elements of the density matrix P_{ab}. The Fock matrix is then constructed by means of equation (2.9), and diagonalized, yielding a set of coefficients $c_{b,\mu}$. For a ground-state calculation, the N lowest-energy atomic orbitals are assumed to be filled, and the density matrix P_{ab} is recalculated. Using the improved density matrix, an improved Fock matrix can be constructed and re-diagonalized, yielding improved coefficients $c_{b,\mu}$, and so on. The procedure is continued until the change produced by successive iterations is very small, at which point the solution is said to be self-consistent.

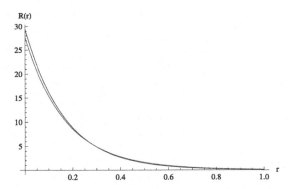

Fig. 2.1 *This figure shows the radial part of the 1s atomic orbital resulting from a Hartree-Fock calculation on carbon. The same orbital without the effects of repulsion is shown for comparison. Since the 1s orbital is an inner-shell orbital, it experiences the almost unscreened effects of the carbon nucleus, but the charge cloud of the orbital is moved outward very slightly. The Hartree-Fock calculation was performed with a radial basis set consisting of 10 Löwdin-orthogonalized Coulomb Sturmians (see Appendix B). The calculation preserved spherical symmetry by using the occupation numbers discussed in the text.*

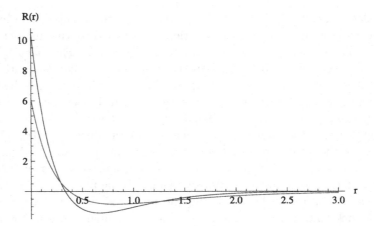

Fig. 2.2   *The radial dependence calculated 2s Hartree-Fock atomic spin-orbitals is shown here. The orbitals are very different from the carbon 2s orbital without the effects on interelectron repulsion, which is shown for comparison.*

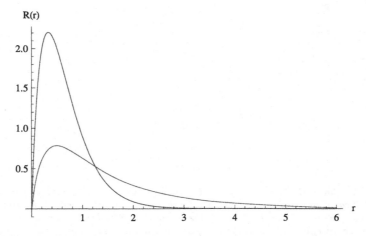

Fig. 2.3   *This figure shows the radial dependence of the 2p spin-orbitals resulting from a Hartree-Fock calculation on the neutral carbon atom. The same orbital without the effects of interelectron repulsion is shown for comparison.*

2.2   Automatic generation of symmetry-adapted configurations

The Hartree-Fock-Roothaan SCF procedure produces a set of orbitals that are optimal (except for errors produced by basis set truncation) when the wave function is required to be a single Slater determinant. However, this requirement is unrealistic. To improve the approximation, one often performs Configuration Interaction (CI) calculations, in which the wave function is allowed to be more general in the sense that it may consist of many Slater determinants, i.e., many configurations. In SCF-CI ground-state calculations, a set of configurations $|\Phi_\nu\rangle$ are constructed from the filled and virtual orbitals of the original SCF calculation by promoting a few electrons from filled orbitals to virtual orbitals. The Hamiltonian based on these configurations is then constructed and diagonalized. The number of configurations in such a CI calculation is often extremely large, and thus it is desirable to find symmetry-adapted basis sets for such a calculation. If the atomic orbitals have the orthonormality property

$$\int d^3x_i \chi_\mu^*(\mathbf{x}_i)\chi_{\mu'}(\mathbf{x}_i) = \delta_{\mu,\mu'} \tag{2.14}$$

then the configurations are also orthonormal:

$$\langle \Phi_{\nu'}|\Phi_\nu\rangle = \delta_{\nu',\nu} \tag{2.15}$$

Substituting the superposition (2.3) into the Schrödinger equation (2.1) and taking the scalar product with a conjugate configuration yields:

$$\sum_{\nu=1}^M \langle \Phi_{\nu'}|\,[H - E_\kappa]\,|\Phi_\nu\rangle C_{\nu,\kappa} = 0 \qquad \nu' = 1,2,3,\ldots,M \tag{2.16}$$

Since the configurations have the orthonormality property (2.15), this yields a set of secular equations of the form

$$\sum_{\nu=1}^M [H_{\nu',\nu} - E_\kappa \delta_{\nu',\nu}]\, C_{\nu,\kappa} = 0 \qquad \nu' = 1,2,3,\ldots,M \tag{2.17}$$

where

$$H_{\nu',\nu} \equiv \langle \Phi_{\nu'}|H|\Phi_\nu\rangle \tag{2.18}$$

The configuration indices ν are sets of atomic orbital indices:

$$\nu = (\mu,\mu',\mu'',\ldots) = (n,l,m,m_s;n',m',l',m_s';\ldots) \tag{2.19}$$

We can notice that the configurations in our basis set are eigenfunction of both the z-component of total orbital angular momentum

$$L_z|\Phi_\nu\rangle = M_L|\Phi_\nu\rangle \qquad M_L = m + m' + m'' + \cdots \tag{2.20}$$

and the z-component of total spin:

$$S_z|\Phi_\nu\rangle = M_S|\Phi_\nu\rangle \qquad M_S = m_s + m_s' + m_s'' + \cdots \tag{2.21}$$

The boundary conditions obeyed by the configurations in our basis set are those requiring finiteness at the origin and square integrability. Since H includes the interelectron repulsion potential, only operations $\mathcal{R}$ that leave the interelectron distances invariant can be included in $\mathcal{G}$. These include rotations of the system as a whole about the position of the nucleus, and rotations in spin space. We can relabel the configurations by a new set of indices which are divided into two categories. The first category of indices, "a", are unaffected by the elements of $\mathcal{G}$, while the indices in the second category, "b", are affected by these operations. Thus our new configuration labels have the form:

$$\nu = (a; b) \tag{2.22}$$

where

$$a = (n, l, n'l', n'', l'', \ldots, M_L, M_S) \qquad b = (m, m_s, m', m'_s, m'', m''_s, \ldots) \tag{2.23}$$

The indices in category "b" are subject to the restrictions:

$$
\begin{aligned}
m + m' + m'' + \cdots &= M_L \\
m_s + m'_s + m''_s + \cdots &= M_S \\
-l \leq m &\leq l \\
-l' \leq m' &\leq l' \\
-l'' \leq m'' &\leq l'' \\
&\vdots \\
m_s &= \pm 1/2 \\
m'_s &= \pm 1/2 \\
&\vdots
\end{aligned}
\tag{2.24}
$$

The indices in category "a" label invariant subsets of our original set of basis functions.

$$W_k = \operatorname{span} \left\{ \left| \Phi_{a_k, b_1} \right\rangle, \left| \Phi_{a_k, b_2} \right\rangle, \left| \Phi_{a_k, b_3} \right\rangle, \ldots \right\} \tag{2.25}$$

2.3   Russell-Saunders states

Russell-Saunders states of atoms and atomic ions are defined to be simultaneous eigenfunctions of the atomic Hamiltonian together with the operators L^2, S^2, L_z and S_z, all of which commute with the Hamiltonian and with each other. Here L^2 and S^2 are the operators that represent total orbital angular momentum and total spin, while L_z and S_z represent their z-components.

We can also look at the Russell-Saunders states from the point of view of group theory. Irreducible representations of the group of rotations about the origin in space and in spin-space, are labeled by the quantum numbers representing total orbital angular momentum and total spin

$$\alpha = (L, S) \tag{2.26}$$

while basis functions within an irreducible representation can be labeled

$$n = (M_L, M_S) \tag{2.27}$$

where M_L is the quantum number representing the z-component of total angular momentum, while the quantum number M_S corresponds to the z-component of total spin. The secular equation for an invariant block is given by

$$\sum_b \left[\langle \Phi_{a,b'} | H | \Phi_{a,b} \rangle - \lambda_\kappa \delta_{b',b} \right] c_{b,\kappa}^a = 0 \tag{2.28}$$

$$| \omega_a^{L,S,M_L,M_S} \rangle = \sum_b | \Phi_{a,b} \rangle c_{b,\kappa}^a \tag{2.29}$$

The resulting eigenfunctions are symmetry-adapted basis functions of the Russell-Saunders type. After their symmetry has been identified by means of a target function, they can be used to construct symmetry-adapted basis sets to represent states of known symmetry

$$| \Psi_\kappa^{L,S,M_L,M_S} \rangle = \sum_a | \omega_a^{L,S,M_L,M_S} \rangle C_{a,\kappa}^{L,S,M_L,M_S} \tag{2.30}$$

by solving the reduced set of secular equations:

$$\sum_a \left[\langle \omega_{a'}^{L,S,M_L,M_S} | H | \omega_a^{L,S,M_L,M_S} \rangle - \lambda_\kappa \delta_{a',a} \right] C_{a,\kappa}^{L,S,M_L,M_S} = 0 \tag{2.31}$$

2.4 Some illustrative examples

Let us consider the carbonlike isoelectronic series, i.e. the series that contains the carbon atom and a number of ions with various nuclear charges Z, but where the number of electrons N is always equal to 6. If interelectron repulsion is entirely neglected, the energy of states in the carbonlike isoelectronic series will be given by

$$E_\nu^0 = -\frac{1}{2} Z^2 \left(\frac{1}{n_1^2} + \frac{1}{n_2^2} + \frac{1}{n_3^2} + \frac{1}{n_4^2} + \frac{1}{n_5^2} + \frac{1}{n_6^2} \right) \tag{2.32}$$

In other words, if interelectron repulsion is entirely neglected, the energies of states for the 6-electron isoelectronic series will be the energies of 6 independent electrons moving in the attractive Coulomb potential of a nucleus with charge Z. Here the index ν stands for the set of quantum numbers $(n_1, l_1, m_1, ms_1; n_2, l_2, m_2, ms_2; \ldots; n_6, l_6, m_6, ms_6)$ labeling the configurations. As Z becomes very large compared with $N = 6$, the Coulomb attraction of the nucleus dominates over interelectron repulsion, and the nonrelativistic energies of the 6-electron isoelectronic series approach E_ν^0. The lowest value of E_ν^0 allowed by the Pauli exclusion principle is

$$E_\nu^0 = -\frac{1}{2}Z^2 \left(\frac{1}{1} + \frac{1}{1} + \frac{1}{4} + \frac{1}{4} + \frac{1}{4} + \frac{1}{4} \right) = -\frac{3}{2}Z^2 \qquad (2.33)$$

This is the state with 2 electrons in the $n = 1$ shell and 4 electrons in the $n = 2$ shell. The number of Pauli-allowed configurations corresponding to this limiting value of E_ν^0 is given by the binomial coefficient

$$\binom{8}{4} = 70 \qquad (2.34)$$

since there is only one way to put 2 electrons into the $n = 1$ shell. The binomial coefficient gives us the number of different ways to put the remaining 4 electrons into the $n = 2$ shell. We can, if we like, use the sets of configurations corresponding to particular values of E_ν^0 as invariant subsets, but equation (2.23) tells us that the invariant subsets can be made smaller. We can base them on subshells rather than shells. If we do this for the case just discussed, we obtain the following smaller invariant subsets corresponding to the distribution of the electrons within subshells:

$$(1s)^2(2s)^2(2p)^2 \qquad \binom{6}{2} = 15$$

$$(1s)^2(2s)(2p)^3 \qquad \binom{2}{1} \times \binom{6}{3} = 2 \times 20 = 40 \qquad (2.35)$$

$$(1s)^2(2p)^4 \qquad \binom{6}{4} = 15$$

The first invariant subset contains 15 configurations, the second 40 configurations and the third 15. Notice that 15+40+15=70.

If we promote one of the electrons from the $n = 1$ shell to the $n = 2$ shell, the E_ν^0 value becomes

$$E_\nu^0 = -\frac{1}{2}Z^2 \left(\frac{1}{1} + \frac{1}{4} + \frac{1}{4} + \frac{1}{4} + \frac{1}{4} + \frac{1}{4} \right) = -\frac{9}{8}Z^2 \qquad (2.36)$$

The number of configurations corresponding to this value of E_ν^0 is

$$\begin{pmatrix} 2 \\ 1 \end{pmatrix} \times \begin{pmatrix} 8 \\ 5 \end{pmatrix} = 2 \times 56 = 112 \qquad (2.37)$$

that is to say, 2 ways of putting 1 electron into the $n = 1$ shell, and 56 ways of putting the remaining 5 electrons into the $n = 2$ shell. This set of configurations could be used as an invariant subset, or it could be split into smaller invariant subsets based on subshells:

$$(1s)(2s)^2(2p)^3 \qquad \begin{pmatrix} 2 \\ 1 \end{pmatrix} \times \begin{pmatrix} 6 \\ 3 \end{pmatrix} = 2 \times 20 = 40$$

$$(1s)(2s)(2p)^4 \qquad \begin{pmatrix} 2 \\ 1 \end{pmatrix} \times \begin{pmatrix} 2 \\ 1 \end{pmatrix} \times \begin{pmatrix} 6 \\ 4 \end{pmatrix} = 2 \times 2 \times 15 = 60 \qquad (2.38)$$

$$(1s)(2p)^5 \qquad \begin{pmatrix} 2 \\ 1 \end{pmatrix} \times \begin{pmatrix} 6 \\ 5 \end{pmatrix} = 2 \times 6 = 12$$

Once again, the sum of the dimensions of the invariant subsets based on subshells is equal to the dimension of the larger invariant subset based on shells: 40+60+12=112.

We can continue in this way and obtain a large number of Russell-Saunders states of the carbon-like atom or ion. The Russell-Saunders eigenfunctions and energies obtained in this way are valuable in themselves, since they give approximate energies and wave functions for the ground state and various excited states of the atom or ion. However, the symmetry-adapted basis functions obtained in this way can together be used as the starting point for a more refined configuration interaction calculation on the system. In such a calculation, only basis functions of the same symmetry will mix together, i.e. configurations will mix together only if they correspond to the same eigenvalues of the L^2, S^2, L_z and S_z. Thus we can concentrate on a particular symmetry, for example ^{3}S states with $M_l = 0$ and $M_S = 1$, and we can perform CI calculation based exclusively on Russell-Saunders states of this symmetry. In this way, we can greatly reduce the dimension of the large representation of H that must be diagonalized.

As a second example, we can consider the 14-electron silicon-like isoelectronic series, for which

$$E_\nu^0 = -\frac{1}{2}Z^2 \left(\sum_{j=1}^{14} \frac{1}{n_j^2} \right) \qquad (2.39)$$

The smallest Pauli-allowed value of E_ν^0 is

$$E_\nu^0 = -\frac{1}{2}Z^2 \left(\frac{2}{1} + \frac{8}{4} + \frac{4}{9}\right) = -\frac{20}{9}Z^2 \tag{2.40}$$

This corresponds to the situation where 2 electrons are in the $n = 1$ shell, with 8 electrons in the $n = 2$ shell and 4 electrons in the $n = 3$ shell. The number of allowed configurations corresponding to this value of E_ν^0 is

$$\binom{18}{4} = 3060 \tag{2.41}$$

which is a moderately large number. However, we can split the block into smaller invariant subsets based on subshells:

$$
\begin{aligned}
(core)(3s)^2(3p)^2 \qquad &\binom{6}{2} = 15 \\[6pt]
(core)(3s)(3p)^3 \qquad &\binom{2}{1} \times \binom{6}{3} = 2 \times 20 = 40 \\[6pt]
(core)(3p)^4 \qquad &\binom{6}{4} = 15 \\[6pt]
(core)(3s)^2(3d)^2 \qquad &\binom{10}{2} = 45 \\[6pt]
(core)(3s)(3d)^3 \qquad &\binom{2}{1} \times \binom{10}{3} = 2 \times 120 = 240 \\[6pt]
(core)(3d)^4 \qquad &\binom{10}{4} = 210
\end{aligned}
\tag{2.42}
$$

$$\vdots \quad \text{etc.} \qquad \vdots$$

with $15 + 40 + 15 + 45 + 240 + 210 + \cdots = 3060$. In this example, for the E_ν^0-block that we are considering, the $n = 1$ and $n = 2$ shells are completely filled, we have 4 electrons that can be partitioned into 3 boxes (the 3s, 3p and 3d subshells), and we must consider all possible partitions.

2.5   The Slater-Condon rules

In order to evaluate the matrix elements of the Hamiltonian between configurations, we need to make use of the Slater-Condon rules. Appendix D discusses the rules in a generalized form that needs to be used when the orbitals making up the configurations are not orthonormal. However, when the orbitals are orthonormal, as they are here, the generalized Slater-Condon rules of Appendix D reduce to the following simplified form: For the matrix element between two identical configurations, we have:

$$\langle \Phi_0 | H | \Phi_0 \rangle = \sum_{\mu=1}^{N} \int d^3 x_1 \chi_\mu^*(\mathbf{x}_1) h^c(\mathbf{x}_1) \chi_\mu(\mathbf{x}_1)$$

$$+ \frac{1}{2} \sum_{\mu=1}^{N} \sum_{\mu'=1}^{N} \int d^3 x_1 \int d^3 x_2 \chi_\mu^*(\mathbf{x}_1) \chi_{\mu'}^*(\mathbf{x}_2) \frac{1}{r_{1,2}} \left(1 - \mathcal{P}_{1,2}\right) \chi_{\mu'}(\mathbf{x}_2) \chi_\mu(\mathbf{x}_1)$$

$$(2.43)$$

Such a matrix element could, for example, represent the energy of the ground state in the Hartree-Fock approximation if the sums are taken over the filled orbitals. Alternatively the formula could represent a diagonal element between configurations containing virtual orbitals, but in that case, the sums need to be taken over those orbitals that are represented in the configuration $|\Phi_0\rangle$.

After the convergence of the Hartree-Fock-Roothaan calculation, we have a set of coefficients $c_{a,\mu}$ which are used to express the filled and virtual orbitals $\chi_\mu(\mathbf{x}_1)$ in terms of the basis functions $u_a(\mathbf{x}_1)$:

$$\chi_\mu(\mathbf{x}_1) = \sum_a u_a(\mathbf{x}_1) c_{a,\mu} \qquad (2.44)$$

Using these coefficients, we can rewrite the matrix element $\langle \Phi_0 | H | \Phi_0 \rangle$ in terms of the density matrix $P_{a,b}$, the core Hamiltonian matrix, $h_{a,b}^c$, and the Fock matrix $F_{a,b}$:

$$\sum_{\mu=1}^{N} \int d^3 x_1 \chi_\mu^*(\mathbf{x}_1) h^c(\mathbf{x}_1) \chi_\mu(\mathbf{x}_1)$$

$$= \sum_a \sum_b \sum_{\mu=1}^{N} c_{a,\mu}^* c_{b,\mu} \int d^3 x_1 u_a^*(\mathbf{x}_1) h^c(\mathbf{x}_1) u_b(\mathbf{x}_1) \qquad (2.45)$$

$$\equiv \sum_{a,b} P_{a,b}\, h_{a,b}^c$$

where

$$P_{a,b} \equiv \sum_{\mu=1}^{N} c_{a,\mu}^* c_{b,\mu} \qquad (2.46)$$

If $|\Phi_0\rangle$ is the Slater determinant representing the Hartree-Fock ground state, the sum runs over the filled orbitals. In the more general case, the sum runs over those orbitals that are contained in $|\Phi_0\rangle$. Treating the Coulomb and exchange term similarly we have:

$$\langle\Phi_0|H|\Phi_0\rangle = \sum_{a,b} P_{a,b}\left(h_{a,b}^c + \frac{1}{2}G_{a,b}\right) \tag{2.47}$$

where

$$G_{a,b} \equiv \sum_{c,d} P_{c,d}\Gamma_{a,b,(c,d)} \tag{2.48}$$

This can be rewritten in terms of the Fock matrix as:

$$\langle\Phi_0|H|\Phi_0\rangle = \frac{1}{2}\sum_{a,b} P_{a,b}\left(h_{a,b}^c + F_{a,b}\right) \tag{2.49}$$

Now consider a configuration $|\Phi_{\mu_i\to\mu_j}\rangle$ that differs from $|\Phi_0\rangle$ by having the orbital χ_{μ_i} replaced by χ_{μ_j}. The second Slater-Condon rule states that

$$\langle\Phi_0|H|\Phi_{\mu_i\to\mu_j}\rangle = \int d^3x_1\,\chi_{\mu_i}^*(\mathbf{x}_1)h^c(\mathbf{x}_1)\chi_{\mu_j}(\mathbf{x}_1)$$

$$+ \frac{1}{2}\sum_{\mu=1}^{N}\int d^3x_1\int d^3x_2\,\chi_{\mu_i}^*(\mathbf{x}_1)\chi_\mu^*(\mathbf{x}_2)\frac{1}{r_{1,2}}\left(1-\mathcal{P}_{1,2}\right)\chi_\mu(\mathbf{x}_2)\chi_{\mu_j}(\mathbf{x}_1)$$

$$\tag{2.50}$$

where the sum runs over those orbitals that are filled in $|\Phi_0\rangle$.

Finally let us consider a configuration $|\Phi_{\mu_i\to\mu_j}^{\mu_k\to\mu_l}\rangle$, which differs from $|\Phi_0\rangle$ by two orbitals. The third Slater-Condon rule states that

$$\langle\Phi_0|H|\Phi_{\mu_i\to\mu_j}^{\mu_k\to\mu_l}\rangle$$
$$= \int d^3x_1\int d^3x_2\,\chi_{\mu_i}^*(\mathbf{x}_1)\chi_{\mu_k}^*(\mathbf{x}_2)\frac{1}{r_{1,2}}\left(1-\mathcal{P}_{1,2}\right)\chi_{\mu_l}(\mathbf{x}_2)\chi_{\mu_j}(\mathbf{x}_1) \tag{2.51}$$

For configurations that differ from $|\Phi_0\rangle$ by three or more orbitals, the matrix element is zero.

2.6 Diagonalization of invariant blocks using the Slater-Condon rules

After performing a Hartree-Fock calculation, we can construct configurations based on the filled and virtual orbitals, and we can divide these configurations into invariant subsets. The representations of the Hamiltonian operator of the system can then be found by means of the Slater-Condon rules. When this invariant block is diagonalized, we obtain Russell-Saunders states, which are symmetry-adapted basis functions that can be used in a more refined configuration interaction calculation aimed at states of a particular symmetry.

As a particular example of this procedure, we can consider the neutral beryllium atom. A Hartree-Fock calculation on this atom, using a radial basis set consisting of 10 Löwdin-orthogonalized Coulomb Sturmians (see Appendix B), yields a HF ground-state energy of -14.572960 Hartrees, which can be compared with NIST's best HF value of -14.572967 Hartrees. The lowest-energy invariant block of configurations contains

$$\binom{8}{2} = 28 \tag{2.52}$$

configurations. In other words, for this block, the n=1 shell is always completely filled. This leaves 2 electrons to go into the 8 available spin-orbitals of the n=2 shell, and there are 28 different Pauli-allowed ways of putting them into the shell. The Russell-Saunders states obtained by diagonalizing this 28×28 dimensional invariant block invariant block are shown in Table 2.1.

For the case of an open-shell system like the neutral carbon atom, Russell-Saunders states can be obtained with little effort if the Hartree-Fock calculation is performed in a symmetrical way. This can be done for carbon by introducing the occupation numbers

$$\mathbf{n} = \left(1, 1, 1, 1, \frac{1}{3}, \frac{1}{3}, \frac{1}{3}, \frac{1}{3}, \frac{1}{3}, \frac{1}{3}\right) \tag{2.53}$$

In this symmetrical model of the neutral carbon atom, the $(1s)^2$ shell and the $(2s)^2$ subshell are visualized as fully occupied, while the spin-orbitals in the $(2p)^6$ subshell are partially occupied, each with occupation number $1/3$. The sum of the occupation numbers is 6, the total number of electrons on neutral carbon. The charge and bond-order matrix is then given by

$$P_{a,b} = \sum_{\mu=1}^{10} c_{a,\mu}^* c_{b,\mu} \mathbf{n}_\mu \tag{2.54}$$

The Hartree-Fock ground state energy obtained by this symmetrical calculation is -37.3407 Hartrees, several tenths of a Hartree above the energy that would be obtained by a calculation aimed specifically at obtaining the unsymmetrical ^{3}P carbon ground state. However the symmetrical calculation leads to Russell-Saunders configurations, and greater accuracy can be obtained by using these as symmetry-adapted basis functions in a large configuration interaction calculation.

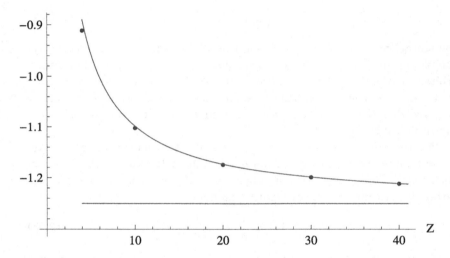

Fig. 2.4   *This figure shows E/Z^2 for the ground state energies of neutral beryllium and for four other ions in the beryllium-like isoelectronic series as functions of Z. The energies (dots) were calculated by diagonalizing the 28×28 block of configurations discussed in the text, and they are compared with the approximate curve $E_a/Z^2 \approx -(Z\sqrt{5/2} - .986712)^2/(2Z^2)$ derived in Chapter 4. The calculated energies approach the approximate curve as Z becomes large, but for the energies to approach the experimental values at large Z, relativistic corrections would be needed. The horizontal line at $E/Z^2 = -5/4$ indicates the ground state electronic energies if interelectron repulsion is entirely neglected.*

Table 2.1: Russell-Saunders states resulting from the diagonalization of the 28×28 block for neutral beryllium

energy	term	degen.	L	S	dominant configuration
−14.5836	^{1}S	1	0	0	$(1s)^2(2s)^2$
−14.4565	^{3}P	9	1	1	$(1s)^2(2s)(2p)$
−14.3798	^{1}P	3	1	0	$(1s)^2(2s)(2p)$
−14.2209	^{3}P	9	1	1	$(1s)^2(2p)^2$
−14.1998	^{1}D	5	2	0	$(1s)^2(2p)^2$
−14.1577	^{1}S	1	0	0	$(1s)^2(2p)^2$

Table 2.2: Russell-Saunders states resulting from the diagonalization of the 28×28 block of configurations for the beryllium-like (4-electron) isoelectronic series. Energies are compared with the approximate expression $E_a \approx -(Z\sqrt{5/2} - |\lambda_\kappa|)^2/2$, which is derived in Chapter 4. The approximate expression becomes increasingly accurate as Z increases, but in order for the calculated values to approach the experimental ones at large Z, relativistic effects would have to be included.

term	$|\lambda_\kappa|$	energies	$Z = 10$	$Z = 20$	$Z = 40$
^{1}S	0.986172	E E/Z^2 E_a/Z^2	-110.26 -1.1026 -1.0989	-469.70 -1.1743 -1.1733	-1938.6 -1.2116 -1.2113
^{3}P	1.02720	E E/Z^2 E_a/Z^2	-109.71 -1.0971 -1.0929	-468.46 -1.1711 -1.1701	-1935.96 -1.2099 -1.2097
^{1}P	1.06426	E E/Z^2 E_a/Z^2	-109.18 -1.0918 -1.0874	-467.34 -1.1683 -1.1673	-1933.7 -1.2083 -1.2076
^{3}P	1.09169	E E/Z^2 E_a/Z^2	-108.89 -1.0889 -1.0833	-466.61 -1.1665 -1.1652	-1932.1 -1.2085 -1.2072
^{1}D	1.10503	E E/Z^2 E_a/Z^2	-108.73 -1.0873 -1.0812	-466.24 -1.1656 -1.1642	-1931.3 -1.2071 -1.2067
^{1}S	1.13246	E E/Z^2 E_a/Z^2	-108.57 -1.0857 -1.0774	-465.87 -1.1647 -1.1621	-1930.5 -1.2066 -1.2056

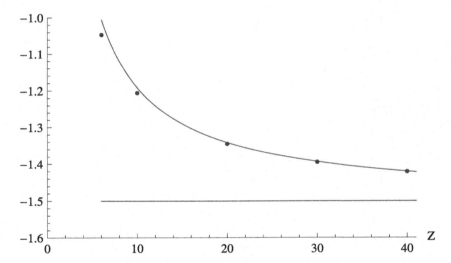

Fig. 2.5   *This figure, which is similar to Figure 2.4, shows E/Z^2 for the ground state energies of neutral carbon and for four other ions in the carbon-like isoelectronic series as functions of Z. The energies (dots) were calculated by diagonalizing the 70×70 block of configurations discussed in the text, and they are compared with the approximate curve $E_a/Z^2 \approx -(Z\sqrt{3} - 1.88151)^2/(2Z^2)$ derived in Chapter 4. The horizontal line at $E/Z^2 = -3/2$ indicates the ground state electronic energies if interelectron repulsion is entirely neglected.*

Chapter 3

EXAMPLES FROM QUANTUM CHEMISTRY

3.1 The Hartree-Fock-Roothaan method applied to molecules

Let us now consider the case of an N-electron molecule. We can try to build up the wave function of the molecule from configurations based on molecular orbitals which are solutions to the Hartree-Fock equations. The Hartree-Fock-Roothaan procedure goes through in the same way as is described in Chapter 2, except that the core Hamiltonian is given by

$$h^c(\mathbf{x}_i) = -\frac{1}{2}\nabla_i^2 - \sum_a \frac{Z_a}{|\mathbf{x}_i - \mathbf{X}_a|} \qquad (3.1)$$

After performing this calculation we are left with a set of filled and virtual molecular spin-orbitals of the form

$$\varphi_\mu(\mathbf{x}_i) = \sum_b u_b(\mathbf{x}_i) c_{b,\mu} \qquad (3.2)$$

where the functions u_a are a set of basis functions, usually Cartesian Gaussians centered on the various atoms of the molecule, multiplied by an α or β spin function. The Hamiltonian of the molecule is

$$H = \sum_{i=1}^N h^c(\mathbf{x}_i) + \sum_{i>j}^N \frac{1}{r_{ij}} \qquad (3.3)$$

The wave functions $|\Psi_\kappa\rangle$ of the molecule are eigenfunctions of the Hamiltonian H:

$$H|\Psi_\kappa\rangle = E_\kappa|\Psi_\kappa\rangle \qquad (3.4)$$

We can approximate the wave function by a linear superposition of configurations:

$$|\Psi_\kappa\rangle = \sum_{\nu=1}^M |\Phi_\nu\rangle C_{\nu,\kappa} \qquad (3.5)$$

Each configuration is a Slater determinant

$$
|\Phi_\nu\rangle = |\varphi_\mu\varphi_{\mu'}\varphi_{\mu''}\cdots| \equiv \frac{1}{\sqrt{N!}}
\begin{vmatrix}
\varphi_\mu(\mathbf{x}_1) & \varphi_{\mu'}(\mathbf{x}_1) & \varphi_{\mu''}(\mathbf{x}_1) & \cdots \\
\varphi_\mu(\mathbf{x}_2) & \varphi_{\mu'}(\mathbf{x}_2) & \varphi_{\mu''}(\mathbf{x}_2) & \cdots \\
\varphi_\mu(\mathbf{x}_3) & \varphi_{\mu'}(\mathbf{x}_3) & \varphi_{\mu''}(\mathbf{x}_3) & \cdots \\
\vdots & \vdots & \vdots &
\end{vmatrix}
\tag{3.6}
$$

The index ν labeling the configurations contains the indices of the molecular orbitals that enter the configuration:

$$
\nu = (\mu, \mu', \mu'', \ldots)
\tag{3.7}
$$

The 1-electron molecular orbitals are orthonormal,

$$
\int d^3 x_i \varphi_{\mu'}^*(\mathbf{x}_i)\varphi_\mu(\mathbf{x}_i) = \delta_{\mu',\mu}
\tag{3.8}
$$

since they are eigenfunctions of a Hermitian operator. From this it follows that the configurations are also orthonormal:

$$
\langle \Phi_{\nu'}|\Phi_\nu\rangle = \delta_{\nu',\nu}
\tag{3.9}
$$

Inserting the linear combination of configurations (3.5) into the Schrödinger equation (3.4), and taking the scalar product with the conjugate configuration $\langle\Phi_{\nu'}|$, we obtain

$$
\sum_{\nu=1}^{M} \langle \Phi_{\nu'}| [H - E_\kappa]|\Phi_\nu\rangle C_{\nu,\kappa} = 0 \qquad \nu' = 1, 2, 3, \ldots, M
\tag{3.10}
$$

Because of the orthonormality of the configurations (3.9), this gives us the set of secular equations:

$$
\sum_{\nu=1}^{M} [H_{\nu',\nu} - E_\kappa \delta_{\nu',\nu}] C_{\nu,\kappa} = 0 \qquad \nu' = 1, 2, 3, \ldots, M
\tag{3.11}
$$

where

$$
H_{\nu',\nu} \equiv \langle \Phi_{\nu'}|H|\Phi_\nu\rangle
\tag{3.12}
$$

3.2   Construction of invariant subsets

We now wish to divide the large set of configurations $|\Phi_\nu\rangle$ into invariant subsets, i.e. into sets that will be invariant under the symmetry group $\mathcal{G}$ of the molecule. The index ν that labels the configurations indicates which molecular spin-orbitals participate in a particular configuration; and each molecular orbital μ can be subdivided into several indices. The first few

Table 3.1: A few of the low-energy molecular orbitals for homonuclear diatomic molecules

μ	$1\sigma_g$	$1\sigma_u^*$	$2\sigma_g$	$2\sigma_u^*$	$1\pi_{u,1}$	$1\pi_{u,-1}$
λ	0	0	0	0	1	-1
inv	+1	-1	+1	-1	-1	-1

molecular orbitals of a homonuclear diatomic molecule are shown in Table 3.1. If the z-axis is taken in the direction of molecular axis, and if the origin is taken to be a point halfway between the two atoms, the symmetry group of a homonuclear diatomic molecule consists of rotations about the z-axis, the identity element and the inversion operator. The orbitals are labeled with a subscript u or g depending on whether they are even or odd under inversion. They are also labeled according to the value of λ, where $L_z\varphi_\mu = \pm\lambda\varphi_\mu$. For $\lambda = 0, 1, 2, \ldots$, the orbitals are labeled respectively with the Greek letters $\sigma, \pi, \delta, \ldots$, the system being analogous to the labeling of atomic orbitals as s,p,d,f,.. according to their l-values.

Now suppose that we wish to divide a large set of configurations based on the molecular orbitals of a homonuclear diatomic molecule into invariant subsets, i.e., into subsets that will be closed under the operations of the symmetry group $\mathcal{G}$ of the molecule. Just as in the case of atoms, we can do this by dividing the sets of indices ν into two categories;

$$\nu = (\mu, \mu', \mu'', \ldots) = (a; b) \tag{3.13}$$

In the category a are those indices that are left invariant by the operations of $\mathcal{G}$. The sets of configurations are even or odd under the operation of the operator that inverts the coordinates of all the electrons in the molecule, so that the even and odd configurations can be placed in different invariant subsets. Finally the eigenvalue of L_z is not affected by the operations of $\mathcal{G}$. Together with the overall parity of the configurations, the eigenvalue value of L_z belongs to category a. Furthermore, the eigenvalue of the z-component of total spin operator belongs to category a. The remaining indices in ν can be placed in category b.

In Chapter 2, we wrote

$$W_k = (|\Phi_{a_k,b_1}\rangle, |\Phi_{a_k,b_2}\rangle, |\Phi_{a_k,b_3}\rangle, \ldots) \tag{3.14}$$

The principle is the same in the case of molecules: The indices in category "a" (those that remain invariant under the operation of $\mathcal{G}$) are the same for all the configurations in a particular invariant subspace. The indices are changed by the elements of $\mathcal{G}$ are in category "b", and they label the different configurations that belong to an invariant subspace. Just as in the case of molecules, S and M_S belong to category "a", while $m_s, m_s', m_s'', \ldots$ belong to category "b". For molecules, the symmetry-names $\gamma, \gamma', \gamma'', \ldots$ of the molecular spin-orbitals belong to category "a", but in the case of many-dimensional irreducible representations, the index that characterizes the particular basis function within an irreducible representation belongs to category "b", and these indices run over all their possible Pauli-allowed values within an invariant subset of configurations as do the values of $m_s, m_s', m_s'', \ldots$.

3.3   The trigonal group $\mathbf{C_{3v}}$; the NH$_3$ molecule

As a second example to illustrate the discussion given above, we can consider the trigonal group C_{3v}, one of the simplest molecular point groups. To make the example more concrete, we can think of the ammonia molecule, NH$_3$, which is carried over into itself by the elements of C_{3v}. In the ammonia molecule, the equilibrium atomic positions are given approximately by

$$\mathbf{X_1} = R\left(\frac{2\sqrt{2}}{3}, 0, -\frac{1}{3}\right)$$

$$\mathbf{X_2} = R\left(-\frac{\sqrt{2}}{3}, \sqrt{\frac{2}{3}}, -\frac{1}{3}\right) \tag{3.15}$$

$$\mathbf{X_3} = R\left(-\frac{\sqrt{2}}{3}, -\sqrt{\frac{2}{3}}, -\frac{1}{3}\right)$$

$$\mathbf{X_4} = (0, 0, 0)$$

where $\mathbf{X_1}$, $\mathbf{X_2}$ and $\mathbf{X_3}$ are the positions of the hydrogens, while $\mathbf{X_4}$ is the position of the nitrogen, at which the origin of the coordinate system is placed. The z-axis is perpendicular to the plane of the three hydrogens. The elements of the point-group C_{3v} are as follows: the identity element E; two rotations about the z-axis $(2C_3)$, through $120°$ and $240°$ respectively; and three reflections $(3\sigma_v)$ through three different planes, each of which includes the z-axis and one of the hydrogen atoms. As can be seen from

Table 3.2: Character table for the group C_{3v}. The representation labeled by E is 2-dimensional.

C_{3v}	E	$2C_3(z)$	$3\sigma_v$	*linear functions and rotations*	*quadratic functions*
A_1	1	1	1	z	$x^2 + y^2, z^2$
A_2	1	1	-1	R_z	
E	2	-1	0	$(1e_x, 1e_y), (-R_y, R_x)$	$(2e_x, 2e_y), (3e_x, 3e_y)$

the character table shown above, C_{3v} has three irreducible representations, A_1, A_2 and E. The first two of these are 1-dimensional representations, while E is 2-dimensional.

In the case of ammonia, only the position of the nitrogen atom is left unchanged by the operations of the group $\mathcal{G} = C_{3v}$, while the hydrogens change places with one another. What are the categories of the configuration index $\nu = (a; b)$ in this case?

Since each of the hydrogen atoms contributes an electron to the system, it now has 10 electrons, and the ground state is a neon-like closed shell. After the Hartree-Fock calculation, the filled spin-orbitals can be denoted by

$$\text{filled}: \quad (1a_{1\uparrow}, 1a_{1\downarrow}, 2a_{1\uparrow}, 2a_{1\downarrow}, 1e_{x\uparrow}, 1e_{y\uparrow}, 1e_{x\downarrow}, 1e_{y\downarrow}, 3a_{1\uparrow}, 3a_{1\downarrow})$$

(3.16)

where the arrow indicates the direction of the spin. We also obtain virtual spin-orbitals:

$$\text{virtual}: \quad (4a_{1\uparrow}, 4a_{1\downarrow}, 5a_{1\uparrow}, 5a_{1\downarrow}, 2e_{x\uparrow}, 2e_{y\uparrow}, 2e_{x\downarrow}, 2e_{y\downarrow}, 6a_{1\uparrow}, 6a_{1\downarrow}, \dots)$$

(3.17)

Let us now consider the Pauli-allowed configurations that can be formed from these filled and virtual orbitals. We can also try to arrange them into subsets that are invariant under the operations of C_{3v}. The neon-like

Hartree-Fock ground-state configuration forms an invariant subset by itself. However, non-trivial subsets occur when we promote electrons from $1e_x$ or $1e_y$ orbitals to higher states.

Suppose that the Hartree-Fock ground state configuration of NH_3 is denoted by $|\Phi_0\rangle$, while the configuration where an electron has been promoted from $1e_{x\uparrow}$ to $2e_{x\uparrow}$ is denoted by $|\Phi_{1e_{x\uparrow}\to 2e_{x\uparrow}}\rangle$. We can now ask what other configurations are generated when the elements of the group C_{3v} act on $|\Phi_{1e_{x\uparrow}\to 2e_{x\uparrow}}\rangle$. The molecular spin-orbitals with a_1 symmetry are unaffected. However, molecular spin-orbitals with e_x symmetry are converted into the corresponding degenerate orbitals with e_y symmetry and vice versa. Thus, the smallest invariant subset of configurations to which this particular excited configuration belongs is:

$$W_0 = \mathrm{span}\left\{\left|\Phi_{1e_{x\uparrow}\to 2e_{x\uparrow}}\right\rangle, \left|\Phi_{1e_{x\uparrow}\to 2e_{y\uparrow}}\right\rangle, \left|\Phi_{1e_{y\uparrow}\to 2e_{x\uparrow}}\right\rangle, \left|\Phi_{1e_{y\uparrow}\to 2e_{y\uparrow}}\right\rangle,\right.$$
$$\left.\left|\Phi_{1e_{x\downarrow}\to 2e_{x\downarrow}}\right\rangle, \left|\Phi_{1e_{x\downarrow}\to 2e_{y\downarrow}}\right\rangle, \left|\Phi_{1e_{y\downarrow}\to 2e_{x\downarrow}}\right\rangle, \left|\Phi_{1e_{y\downarrow}\to 2e_{y\downarrow}}\right\rangle\right\}$$
(3.18)

A remark must be made here concerning the spins: Obviously, the elements of C_{3v} do not mix the spins of the individual molecular spin-orbitals. But rotations in spin-space do so, and these rotations are also part of the symmetry group of NH_3. However, rotations in spin-space preserve total spin S^2 and its z-component, S_z. The invariant subset of configurations shown above corresponds to $M_S = 0$. Two closely-related invariant subsets, corresponding respectively to $M_S = 1$ and $M_S = -1$ are the following:

$$W_1 = \mathrm{span}\left\{\left|\Phi_{1e_{x\downarrow}\to 2e_{x\uparrow}}\right\rangle, \left|\Phi_{1e_{x\downarrow}\to 2e_{y\uparrow}}\right\rangle, \left|\Phi_{1e_{y\downarrow}\to 2e_{x\uparrow}}\right\rangle, \left|\Phi_{1e_{y\downarrow}\to 2e_{y\uparrow}}\right\rangle\right\}$$
(3.19)

and

$$W_{-1} = \mathrm{span}\left\{\left|\Phi_{1e_{x\uparrow}\to 2e_{x\downarrow}}\right\rangle, \left|\Phi_{1e_{x\uparrow}\to 2e_{y\downarrow}}\right\rangle, \left|\Phi_{1e_{y\uparrow}\to 2e_{x\downarrow}}\right\rangle, \left|\Phi_{1e_{y\uparrow}\to 2e_{y\downarrow}}\right\rangle\right\}$$
(3.20)

If a 16×16-dimensional block of the Hamiltonian based on all 16 configurations is diagonalized, the eigenvalues corresponding to $M_S = 1$, $M_S = 0$ and $M_s = -1$ will separate automatically.

Chapter 4

GENERALIZED STURMIANS APPLIED TO ATOMS

4.1   Goscinskian configurations

The Generalized Sturmian Method (Appendix B) is a newly-developed direct method for performing Configuration Interaction calculations on bound states [Avery, 1989]-[Avery and Avery, 2006]. It avoids the initial Hartree-Fock-Roothaan SCF calculation, and it is especially suitable for calculating large numbers of excited states of few-electron atoms or ions.

When the Generalized Sturmian Method is applied to atoms or atomic ions, it is convenient to use basis functions that are Slater determinants:

$$
|\Phi_\nu\rangle = |\chi_\mu \chi_{\mu'} \chi_{\mu''} \cdots | \equiv \frac{1}{\sqrt{N!}}
\begin{vmatrix}
\chi_\mu(\mathbf{x}_1) & \chi_{\mu'}(\mathbf{x}_1) & \chi_{\mu''}(\mathbf{x}_1) & \cdots \\
\chi_\mu(\mathbf{x}_2) & \chi_{\mu'}(\mathbf{x}_2) & \chi_{\mu''}(\mathbf{x}_2) & \cdots \\
\chi_\mu(\mathbf{x}_3) & \chi_{\mu'}(\mathbf{x}_3) & \chi_{\mu''}(\mathbf{x}_3) & \cdots \\
\vdots & \vdots & \vdots &
\end{vmatrix}
\tag{4.1}
$$

built from hydrogenlike atomic spin-orbitals of the form

$$
\chi_\mu(\mathbf{x}_i) \equiv \chi_{n,l,m,m_s}(\mathbf{x}_i) \equiv R_{n,l}(r_i) Y_{l,m}(\theta_i, \phi_i)
\begin{cases}
\alpha_i & m_s = 1/2 \\
\beta_i & m_s = -1/2
\end{cases}
\tag{4.2}
$$

with weighted nuclear charges Q_ν. In other words, the atomic spin-orbitals have the form shown in equation (2.5), with radial functions given by

$$
R_{1,0}(r) = 2Q_\nu^{3/2} e^{-Q_\nu r}
$$

$$
R_{2,0}(r) = \frac{Q_\nu^{3/2}}{\sqrt{2}} \left(1 - \frac{Q_\nu r}{2} \right) e^{-Q_\nu r/2}
$$

$$
R_{2,1}(r) = \frac{Q_\nu^{5/2}}{2\sqrt{6}} \, r \, e^{-Q_\nu r/2}
$$

$$R_{3,0}(r) = \frac{2Q_\nu^{3/2}}{3\sqrt{3}} \left(1 - \frac{2Q_\nu r}{3} + \frac{2Q_\nu^2 r^2}{27}\right) e^{-Q_\nu r/3} \tag{4.3}$$

$$\vdots \qquad \vdots$$

The reader will recognize these as the familiar hydrogenlike radial functions with the nuclear charge Z replaced by Q_ν. If the effective charges Q_ν characterizing the configurations $|\Phi_\nu\rangle$ are chosen in such a way that

$$Q_\nu = \beta_\nu Z = \left(\frac{-2E_\kappa}{\frac{1}{n^2} + \frac{1}{n'^2} + \frac{1}{n''^2} + \cdots}\right)^{1/2} \tag{4.4}$$

so that

$$E_\kappa = -\frac{Q_\nu^2}{2}\left(\frac{1}{n^2} + \frac{1}{n'^2} + \frac{1}{n'^2} + \cdots\right) \tag{4.5}$$

the configurations will obey the approximate N-electron Schrödinger equation:

$$\left[-\frac{1}{2}\sum_{j=1}^{N}\nabla_j^2 + \beta_\nu V_0(\mathbf{x}) - E_\kappa\right]|\Phi_\nu\rangle = 0 \tag{4.6}$$

where

$$V_0(\mathbf{x}) = -\sum_{j=1}^{N}\frac{Z}{r_j} \tag{4.7}$$

is the nuclear attraction potential. In equation (4.6), the energy E_κ is kept constant for the whole basis set, while the weighting factors β_ν are adjusted to make the basis set isoenergetic. Thus the weighting factors β_ν play the role of eigenvalues in equation (4.6). This type of problem has been called the *conjugate eigenvalue problem* by Coulson, Josephs, Goscinski and others [Goscinski, 1968, 2003], and it is characteristic for the equations defining generalized Sturmian basis sets (Appendix B).

To see that with the special choice of weighted charges shown in equation (4.4) $|\Phi_\nu\rangle$ will satisfy (4.6), we first notice that the hydrogenlike atomic orbitals with weighted nuclear charges obey the 1-electron Schrödinger equation:

$$\left[-\frac{1}{2}\nabla_j^2 + \frac{Q_\nu^2}{2n^2} - \frac{Q_\nu}{r_j}\right]\chi_\mu(\mathbf{x_j}) = 0 \tag{4.8}$$

Since the Slater determinant $|\Phi_\nu\rangle$ is an antisymmetrized product of atomic orbitals, all of which obey (4.8), it follows that

$$\left[-\frac{1}{2}\sum_{j=1}^{N}\nabla_j^2\right]|\Phi_\nu\rangle = \left[-\left(\frac{Q_\nu^2}{2n^2} + \frac{Q_\nu^2}{2n'^2} + \cdots\right) + \left(\frac{Q_\nu}{r_1} + \frac{Q_\nu}{r_2} + \cdots\right)\right]|\Phi_\nu\rangle$$

$$= [E_\kappa - \beta_\nu V_0(\mathbf{x})]|\Phi_\nu\rangle \tag{4.9}$$

and thus equation (4.6) is satisfied. Each configuration $|\Phi_\nu\rangle$ has its own effective nuclear charge Q_ν. Within a particular configuration, the hydrogenlike atomic orbitals are orthonormal

$$\int d\tau_j \, \chi_{\mu'}^*(\mathbf{x}_j)\chi_\mu(\mathbf{x}_j) = \delta_{\mu',\mu} \tag{4.10}$$

and they also obey the virial relationship

$$-\int d\tau_j \, |\chi_\mu(\mathbf{x}_j)|^2 \frac{Q_\nu}{r_j} = -\frac{Q_\nu^2}{n^2} \tag{4.11}$$

From equations (4.6), (4.10) and (4.11), it can be shown [Avery, 2000], [Avery and Avery, 2006] that the generalized Sturmian configurations $|\Phi_\nu\rangle$ obey the potential-weighted orthonormality relation

$$\langle \Phi_{\nu'}^* |V_0| \Phi_\nu \rangle = \delta_{\nu',\nu} \frac{2E_\kappa}{\beta_\nu} \tag{4.12}$$

We next introduce the definitions

$$p_\kappa \equiv \sqrt{-2E_\kappa} \tag{4.13}$$

and

$$\mathcal{R}_\nu \equiv \sqrt{\frac{1}{n^2} + \frac{1}{n'^2} + \cdots} \tag{4.14}$$

With the help of these definitions, equation (4.4) can be written in the form

$$Q_\nu = \beta_\nu Z = \frac{p_\kappa}{\mathcal{R}_\nu} \tag{4.15}$$

The set of Sturmian configurations forms a set of isoenergetic solutions of the approximate Schrödinger equation (4.6), where the potential is weighted, and the weighting factors β_ν are chosen in such a way as to insure that all the solutions correspond to a common energy. From (4.13) we can see that their common energy E_κ is related to p_κ by

$$E_\kappa = -\frac{p_\kappa^2}{2} \tag{4.16}$$

In previous publications we have called such atomic configurations *Goscinskian configurations* to recognize Prof. Osvaldo Goscinski's pioneering work in generalizing the concept of Sturmian basis sets [Goscinski, 1968, 2003]. The non-relativistic Schrödinger equation of an N-electron atom has the form:

$$\left[-\frac{1}{2} \sum_{j=1}^{N} \nabla_j^2 + V(\mathbf{x}) - E_\kappa \right] |\Psi_\kappa\rangle = 0 \tag{4.17}$$

where

$$V(\mathbf{x}) = V_0(\mathbf{x}) + V'(\mathbf{x}) \qquad (4.18)$$

Here $V_0(\mathbf{x})$ is the nuclear attraction potential shown in equation (4.7) while $V'(\mathbf{x})$ is the interelectron repulsion potential

$$V'(\mathbf{x}) = \sum_{i>j}^{N} \frac{1}{r_{ij}} \qquad (4.19)$$

We can try to build up the wave function from a superposition of Goscinskian configurations, i.e. from a superposition of isoenergetic solutions of the approximate wave equation (4.6), where V_0 is the nuclear attraction potential of the atom. Thus we write:

$$|\Psi_\kappa\rangle \approx \sum_\nu |\Phi_\nu\rangle C_{\nu,\kappa} \qquad (4.20)$$

Inserting this superposition into (4.17) we have

$$\sum_\nu \left[-\frac{1}{2}\Delta + V(\mathbf{x}) - E_\kappa \right] |\Phi_\nu\rangle C_{\nu,\kappa} \approx 0 \qquad (4.21)$$

However, each of the basis functions obeys (4.6), and therefore we can rewrite (4.21) in the form

$$\sum_\nu \left[V(\mathbf{x}) - \beta_\nu V_0(\mathbf{x}) \right] |\Phi_\nu\rangle C_{\nu,\kappa} \approx 0 \qquad (4.22)$$

The energy term E_κ is now nowhere to be seen, and a remark is perhaps needed here to explain what has happened to it: The configurations in our Generalized Sturmian basis set are isoenergetic. They all correspond to the same energy, E_κ, since the weighting factors β_ν are chosen especially to make them do so. What we have done in going from (4.21) to (4.22) is to choose this energy to be the same as that which appears in (4.21). In other words, the energy to which all the members of our basis set correspond is chosen to be equal to the energy of the state that we are trying to approximate.

If we take the scalar product of (4.22) with a conjugate function from our basis set, we obtain the set of secular equations:

$$\sum_\nu \langle \Phi_{\nu'} | \left[V(\mathbf{x}) - \beta_\nu V_0(\mathbf{x}) \right] |\Phi_\nu\rangle C_{\nu,\kappa} = 0 \qquad (4.23)$$

We now introduce the definitions:

$$T_{\nu',\nu}^0 \equiv -\frac{1}{p_\kappa} \langle \Phi_{\nu'}^* | V_0 | \Phi_\nu \rangle \qquad (4.24)$$

and

$$T'_{\nu',\nu} \equiv -\frac{1}{p_\kappa} \langle \Phi^*_{\nu'} | V' | \Phi_\nu \rangle \qquad (4.25)$$

From the potential-weighted orthonormality relations (4.12) we can see that

$$T^0_{\nu',\nu} = \delta_{\nu'\nu} Z \mathcal{R}_\nu \qquad (4.26)$$

Notice that the nuclear attraction matrix $T^0_{\nu',\nu}$ is both diagonal and energy-independent. The interelectron repulsion matrix $T'_{\nu',\nu}$ can be evaluated using methods discussed in Appendix D, and it is also energy-independent. In order to see that $T'_{\nu',\nu}$ really is energy-independent, we notice that it is built up from terms of the form

$$\frac{1}{p_\kappa} J_{\mu_1,\mu_2,\mu_3,\mu_4} = \frac{1}{p_\kappa} \int d^3x \int d^3x' \; \rho_{\mu_1,\mu_2}(\mathbf{x}) \frac{1}{|\mathbf{x}-\mathbf{x}'|} \rho_{\mu_3,\mu_4}(\mathbf{x}') \qquad (4.27)$$

where densities are defined by

$$\begin{aligned} \rho_{\mu_1,\mu_2}(\mathbf{x}) &\equiv \chi^*_{\mu_1}(\mathbf{x})\chi_{\mu_2}(\mathbf{x}) \\ \rho_{\mu_3,\mu_4}(\mathbf{x}') &\equiv \chi^*_{\mu_3}(\mathbf{x}')\chi_{\mu_4}(\mathbf{x}') \end{aligned} \qquad (4.28)$$

and where the orbitals are the hydrogenlike orbitals with weighted nuclear charge shown in equations (4.2) and (4.3). We now let

$$\begin{aligned} \mathbf{s} &\equiv p_\kappa \mathbf{x} \\ \mathbf{s}' &\equiv p_\kappa \mathbf{x}' \end{aligned} \qquad (4.29)$$

Then, making the substitution $Q_\nu \to p_\kappa/R_\nu$ in (4.3) we have

$$\begin{aligned} \rho_{\mu_1,\mu_2}(\mathbf{x}) &= p_\kappa^3 \tilde{\rho}_{\mu_1,\mu_2}(\mathbf{s}) \\ \rho_{\mu_3,\mu_4}(\mathbf{x}') &= p_\kappa^3 \tilde{\rho}_{\mu_3,\mu_4}(\mathbf{s}') \end{aligned}$$

where $\tilde{\rho}_{\mu_1,\mu_2}(\mathbf{s})$ and $\tilde{\rho}_{\mu_3,\mu_4}(\mathbf{s}')$ are pure functions of $\mathbf{s}$ and $\mathbf{s}'$ respectively. Finally, noticing that

$$\frac{1}{p_\kappa|\mathbf{x}-\mathbf{x}'|} = \frac{1}{|\mathbf{s}-\mathbf{s}'|} \qquad (4.30)$$

we can write

$$\frac{1}{p_\kappa} J_{\mu_1,\mu_2,\mu_3,\mu_4} = \int d^3s \int d^3s' \; \tilde{\rho}_{\mu_1,\mu_2}(\mathbf{s}) \frac{1}{|\mathbf{s}-\mathbf{s}'|} \tilde{\rho}_{\mu_3,\mu_4}(\mathbf{s}') \qquad (4.31)$$

Since the building-blocks from which it composed are independent of p_κ, the interelectron repulsion matrix $T'_{\nu',\nu}$ is also independent of p_κ and hence independent of energy. The energy-independent interelectron repulsion matrix $T'_{\nu',\nu}$ consists of pure numbers (in atomic units) which can be evaluated once and for all and stored.

With the help of equations (4.24)-(4.26), the secular equation (4.23) can be rewritten in the form:

$$\sum_{\nu} \left[-p_\kappa \delta_{\nu',\nu} Z\mathcal{R}_\nu - p_\kappa T'_{\nu',\nu} + \beta_\nu p_\kappa \delta_{\nu',\nu} Z\mathcal{R}_\nu \right] C_{\nu,\kappa} = 0 \qquad (4.32)$$

Finally, using the relationship

$$\beta_\nu Z\mathcal{R}_\nu = p_\kappa \qquad (4.33)$$

and dividing by p_κ, and reversing the signs, we obtain

$$\sum_{\nu} \left[\delta_{\nu',\nu} Z\mathcal{R}_\nu + T'_{\nu',\nu} - p_\kappa \delta_{\nu',\nu} \right] C_{\nu,\kappa} = 0 \qquad (4.34)$$

The Generalized Sturmian secular equation for atoms and atomics ions (4.34) differs in several remarkable ways from the secular equations that would be obtained using a Hamiltonian method:

(1) The kinetic energy term has disappeared.
(2) The nuclear attraction term, $\delta_{\nu',\nu} Z\mathcal{R}_\nu$, is diagonal.
(3) The interelectron repulsion matrix $T'_{\nu',\nu}$ is energy-independent. It consists of dimensionless pure numbers.
(4) Finally, the roots of the secular equations are not energies but values of the parameter p_κ, which is related to the energy spectrum through equation (4.16). The parameter $p_\kappa = \beta_\nu Z\mathcal{R}_\nu = Q_\nu \mathcal{R}_\nu$ can be thought of as a scaling parameter, since the effective nuclear charges associated with the Goscinskian configurations are proportional to it.
(5) The configurations $|\Phi_\nu\rangle$ in the basis set are not fully determined until the secular equations have been solved. Only the form of the basis functions is known in advance, but not the scale. When the secular equation is solved, the resulting spectrum of p_κ values yields not only a spectrum of energies but a nearly optimum set of basis functions for the representation of each state. The basis set for the representation of highly excited states is diffuse, while the set for representation of tightly-bound states is contracted. The step of optimizing Slater exponents for each problem is thus not needed.
(6) Once the energy-independent interelectron repulsion matrix $T'_{\nu',\nu}$ has been constructed, the properties of an entire isoelectronic series can be calculated with almost no additional effort.

4.2 Relativistic corrections

If the number of electrons N is kept constant while Z is allowed to increase, the energies calculated from the Generalized Sturmian secular equation approach those found by solution of the non-relativistic Schrödinger equation, but a relativistic correction must be added in order for the energies to approach experimental values. A crude relativistic correction can be found for a multiconfigurational state $\Psi_\kappa(\mathbf{x}) = \sum_\nu \Phi_\nu(\mathbf{x})C_{\nu\kappa}$ by calculating the ratio of the relativistic energy of the with interelectron repulsion entirely neglected to the non-relativistic energy, again with interelectron repulsion entirely neglected. The ratio can be written in the form

$$f_\kappa(Z) = \frac{E_{\kappa,\text{rel}}}{E_{\kappa,\text{nonrel}}} = \frac{\sum_\nu C_{\nu\kappa}^2 \langle \Phi_\nu | H_0 | \Phi_\nu \rangle_{\text{rel}}}{-\frac{1}{2}Z^2 \sum_\nu C_{\nu\kappa}^2 \mathcal{R}_\nu^2} \tag{4.35}$$

Here

$$\langle \Phi_\nu | H_0 | \Phi_\nu \rangle_{\text{rel}} = \sum_{\mu \in \nu} \epsilon_{\mu,\text{rel}} \qquad \mu = (n,l,m,m_s) \tag{4.36}$$

is the relativistic energy of the configuration $\Phi_\nu(\mathbf{x})$ with interelectron repulsion entirely neglected, while

$$-\sum_{\mu \in \nu} \frac{1}{2}\frac{Z^2}{n^2} = -\frac{1}{2}Z^2 R_\nu^2 \qquad \mu = (n,l,m,m_s) \tag{4.37}$$

is the nonrelativistic energy of $\Phi_\nu(\mathbf{x})$. The quantity $\epsilon_{\mu,\text{rel}}$ represents the relativistic energy of a single electron moving in the attractive Coulomb potential of a nucleus with charge Z. This energy is easy to calculate exactly [Akhiezer and Berestetskii, 1965], if effects such as vacuum polarization and the Lamb shift are neglected. It is given by:

$$\epsilon_{\mu,\text{rel}} = \frac{c^2}{\left[1 + \left(\frac{Z}{c(\gamma+n-|j+1/2|)}\right)^2\right]^{1/2}} - c^2 \tag{4.38}$$

$$\gamma \equiv \sqrt{\left(j+\frac{1}{2}\right)^2 - \left(\frac{Z}{c}\right)^2} \qquad c = 137.036 \tag{4.39}$$

where j is the total angular momentum (orbital plus spin) of a single electron, i.e. $l \pm \frac{1}{2}$. The corrected energy, $f_\kappa(Z)E_{\kappa,\text{nonrel}}$, agrees closely with the experimental values of energies, especially when Z is large compared with N.

The approximate relativistic correction discussed here is by no means confined to the Generalized Sturmian Method. It can be used in quantum calculations of every kind, performed on atoms and molecules. The assumption behind the correction is that relativistic effects are due mainly to the nuclear attraction part of the Hamiltonian, and only to a lesser extent to interelectron repulsion terms.

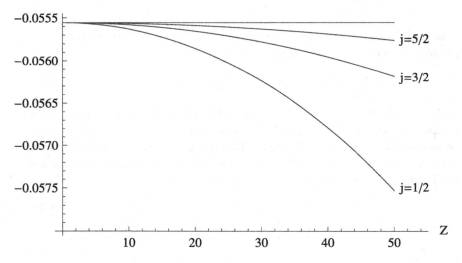

Fig. 4.1   *This figure shows ϵ_μ/Z^2 for n=3. The non-relativistic energy is the horizontal line $\epsilon_{nonrel} = -1/18$, while the relativistic energies are shown by curves.*

Table 4.1: This table shows the relativistic correction for a single electron moving in the field of a nucleus with charge Z, i.e. the relativistic energy without the rest energy, divided by the non-relativistic energy. It is interesting to notice that the correction affects the 4th significant figure of the energy for values of Z as low as 10. In all cases the effect of the relativistic correction is to increase the binding energy.

n	j	$Z=1$	$Z=10$	$Z=20$	$Z=30$
1	$\frac{1}{2}$	1.00001	1.00133	1.00538	1.01228
2	$\frac{1}{2}$	1.00002	1.00167	1.00673	1.01537
2	$\frac{3}{2}$	1.00000	1.00033	1.00133	1.00301
3	$\frac{1}{2}$	1.00001	1.00133	1.00538	1.01226
3	$\frac{3}{2}$	1.00000	1.00044	1.00178	1.00402
3	$\frac{5}{2}$	1.00000	1.00015	1.00059	1.00133

4.3 The large-Z approximation: Restriction of the basis set to an $\mathcal{R}$-block

If interelectron repulsion is entirely neglected, i.e. when disregarding the second term in Eq. (4.34), the calculated energies E_κ become those of a set of N completely independent electrons moving in the field of the bare nucleus:

$$E_\kappa = -\frac{p_\kappa^2}{2} \longrightarrow -\frac{1}{2}Z^2\mathcal{R}_\nu{}^2 = -\frac{Z^2}{2n_1^2} - \frac{Z^2}{2n_2^2} - \cdots - \frac{Z^2}{2n_N^2} \qquad (4.40)$$

In the large-Z approximation, we do not neglect interelectron repulsion, but we restrict the basis set to those Goscinskian configurations that would be degenerate if interelectron repulsion were entirely neglected, i.e., we restrict the basis to a set of configurations all of which correspond to the same value of $\mathcal{R}_\nu$. In that case, the first term in (4.34) is a multiple of the identity matrix, and the eigenvectors $C_{\nu\kappa}$ are the same as those that would be obtained by diagonalizing the energy-independent interelectron repulsion matrix $T'_{\nu'\nu}$, since the eigenfunctions of any matrix are unchanged by adding a multiple of the unit matrix. The simplified secular equation then becomes:

$$\sum_\nu \left[T'_{\nu'\nu} - \lambda_\kappa \delta_{\nu'\nu}\right] C_{\nu\kappa} = 0 \qquad (4.41)$$

The roots are shifted by an amount equal to the constant by which the identity matrix is multiplied:

$$p_\kappa = Z\mathcal{R}_\nu + \lambda_\kappa = Z\mathcal{R}_\nu - |\lambda_\kappa| \qquad (4.42)$$

and the energies become

$$E_\kappa = -\frac{1}{2}(Z\mathcal{R}_\nu - |\lambda_\kappa|)^2 \qquad (4.43)$$

With the relativistic correction of equation (4.35), this becomes

$$E_\kappa = -f(Z)\frac{1}{2}(Z\mathcal{R}_\nu - |\lambda_\kappa|)^2 \qquad (4.44)$$

Since the roots λ_κ are always negative, we may use the form $-|\lambda_\kappa|$ in place of λ_κ to make explicit the fact that interelectron repulsion reduces the binding energies, as of course it must. The roots λ_κ are pure numbers that can be calculated once and for all and stored. From these roots, a great deal of information about atomic states can be found with very little effort.

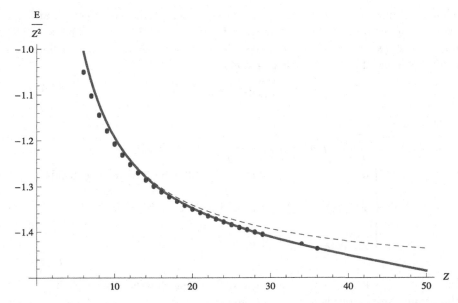

Fig. 4.2 The ground state of the carbon-like isoelectronic series, calculated in the large-Z approximation. The energies divided by Z^2 are shown as functions of Z. Experimental values are indicated by dots, while the energies calculated from equation (4.43) are shown as curves. The lower (solid) curve, which approaches the experimental values with increasing Z, has been corrected for relativistic effects. The upper (dashed) curve is uncorrected.

4.4 Electronic potential at the nucleus in the large-Z approximation

The electronic potential $\varphi(\mathbf{x}_1)$ is related to the electronic density distribution by

$$\varphi(\mathbf{x}_1) = \int d^3 x_1' \frac{\rho(\mathbf{x}_1')}{|\mathbf{x}_1 - \mathbf{x}_1'|} \tag{4.45}$$

If the coordinate system is centered on the nucleus, the electronic potential at the nucleus is then given by

$$\varphi(0) = \int d^3 x_1' \frac{\rho(\mathbf{x}_1')}{|\mathbf{x}_1'|} \tag{4.46}$$

But the electron density corresponding to the state Ψ_κ is defined as

$$\rho(\mathbf{x}_1) = N \int ds_1 \int d^3 x_2 \int ds_2 \cdots \int d^3 x_N \int ds_N \Psi_\kappa^*(\mathbf{x}) \Psi_\kappa(\mathbf{x}) \tag{4.47}$$

where the integral is taken over the spin coordinate of the first electron and over the space and spin coordinates of all the other electrons. The wave function $\Psi_\kappa(\mathbf{x}) = \sum_\nu \Phi_\kappa(\mathbf{x}) B_{\nu\kappa}$ is a linear combination of Goscinskian configurations. Thus the density is given by

$$\rho(\mathbf{x}_1) = \sum_{\nu',\nu} \rho_{\nu'\nu}(\mathbf{x}_1) B^*_{\nu\kappa} B_{\nu\kappa} \qquad (4.48)$$

where

$$\rho_{\nu'\nu}(\mathbf{x}_1) = N \int ds_1 \int d^3x_2 \int ds_2 \cdots \int d^3x_N \int ds_N \Phi^*_{\nu'}(\mathbf{x}) \Phi_\nu(\mathbf{x})$$

$$= \begin{cases} 0 & \text{if } \nu' \text{ and } \nu \text{ differ by 2 or more orbitals} \\ \chi^*_{\mu'}(\mathbf{x}_1)\chi_\mu(\mathbf{x}_1) & \text{if } \nu' \text{ and } \nu \text{ differ only by } \mu \to \mu' \\ \sum_{i=1}^N |\chi_{\mu_i}(\mathbf{x}_1)|^2 & \text{if } \nu' = \nu \end{cases}$$

$$(4.49)$$

In equation (4.49) we have made use of the fact that within an $\mathcal{R}$-block, the atomic spin-orbitals are orthonormal.

Within the framework of the large-Z approximation we have

$$\int dx\ \Psi^*_\kappa(\mathbf{x}) V_0(\mathbf{x}) \Psi_\kappa(\mathbf{x}) = \sum_{\nu'} \sum_\nu B^*_{\nu'\kappa} B_{\nu\kappa} \int dx\ \Phi^*_{\nu'}(\mathbf{x}) V_0(\mathbf{x}) \Phi_\nu(\mathbf{x})$$

$$= -\frac{p_\kappa^2}{\beta_\nu} \sum_\nu |B_{\nu\kappa}|^2$$

$$(4.50)$$

In the second step above, we make use of the potential weighted orthonormality relation (4.12). Further, since $\sum_\nu |B_{\nu\kappa}|^2 = 1$, equation (4.50) reduces to

$$\int dx\ \Psi^*_\kappa(\mathbf{x}) V_0(\mathbf{x}) \Psi_\kappa(\mathbf{x}) = -\frac{p_\kappa^2}{\beta_\nu} = -p_\kappa Z \mathcal{R}_\nu \qquad (4.51)$$

This result can be used to express the electronic potential at the nucleus in a very simple form. Combining (4.46) and (4.47), we obtain

$$\varphi(0) = N \int dx\ \frac{1}{|\mathbf{x}_1|} \Psi^*_\kappa(\mathbf{x}) \Psi_\kappa(\mathbf{x}) \qquad (4.52)$$

From the definition of V_0, equation (4.7), and from the fact that each term in the sum in (4.7) gives the same contribution, we have

$$\varphi(0) = -\frac{1}{Z} \int dx\ \Psi^*_\kappa(\mathbf{x}) V_0(\mathbf{x}) \Psi_\kappa(\mathbf{x}) \qquad (4.53)$$

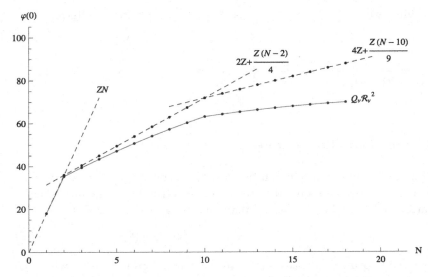

Fig. 4.3 When interelectron repulsion is entirely neglected, the electronic potential at the nucleus is given by $Z\mathcal{R}_\nu{}^2$, which is exactly piecewise linear in N. The effect of interelectron repulsion is to decrease $\varphi(0)$ and to make the dependence only approximately piecewise linear. The figure shows $\varphi(0)$ neglecting interelectron repulsion (upper values) and including it (lower values). The dots are calculated from the electronic densities of the ground state wave functions, whereas the lines are the closed form expressions found in equations (4.57) and (4.55).

Combining equations (4.53) and (4.51) we obtain the extremely simple result:

$$\varphi(0) = p_\kappa \mathcal{R}_\nu \qquad (4.54)$$

which can alternatively be written in the form:

$$\varphi(0) = Z\mathcal{R}_\nu{}^2 - |\lambda_\kappa|\mathcal{R}_\nu \qquad (4.55)$$

or in a third form:

$$\varphi(0) = Q_\nu \mathcal{R}_\nu{}^2 \qquad (4.56)$$

since $Q_\nu = Z - |\lambda_\kappa|/\mathcal{R}_\nu$. From equations (4.54)-(4.56) it follows that for an isonuclear series, the electronic potential at the nucleus depends on N in an approximately piecewise linear way. For example, let us consider the isonuclear series where $Z = 18$. Keeping the nuclear charge Z constant at this value, we begin to add electrons. For the ground state we have:

$$
\mathcal{R}_\nu{}^2 \equiv \frac{1}{n_1^2} + \frac{1}{n_2^2} + \cdots + \frac{1}{n_N^2} = \begin{cases} \frac{N}{1} & N \leq 2 \\[2mm] \frac{2}{1} + \frac{N-2}{4} & 2 \leq N \leq 10 \\[2mm] \frac{2}{1} + \frac{8}{4} + \frac{N-10}{9} & 10 \leq N \leq 18 \end{cases} \quad (4.57)
$$

4.5   Core ionization energies

The large-Z approximation can be used to calculate core-ionization energies, i.e. the energies required to remove an electron from the inner shell of an atom. From (4.43) we can see that this energy will be given by

$$
\Delta E = \frac{1}{2} \left[(Z\mathcal{R}_\nu - |\lambda_\kappa|)^2 - (Z\mathcal{R}_\nu{}' - |\lambda_\kappa'|)^2 \right] \quad (4.58)
$$

where the unprimed quantities refer to the original ground state, while the primed quantities refer to the core-ionized states. Since

$$
\mathcal{R}_\nu{}^2 - \mathcal{R}_\nu{}'^2 = 1 \quad (4.59)
$$

Equation (4.58) can be written in the form

$$
\Delta E - \frac{Z^2}{2} = Z \left[\mathcal{R}_\nu{}'|\lambda_\kappa'| - \mathcal{R}_\nu|\lambda_\kappa| \right] + \frac{|\lambda_\kappa|^2 - |\lambda_\kappa'|^2}{2} \quad (4.60)
$$

Thus we can see that within the framework of the large-Z approximation, the quantity $\Delta E - Z^2/2$ is linear in Z for an isoelectronic series. This quantity represents the contribution of interelectron repulsion to the core ionization energy, since if interelectron repulsion is completely neglected, the core ionization energy is given by $\Delta E = Z^2/2$. Core ionization energies calculated from equations (4.58)-(4.60) are shown in Figures 4.4 through 4.6.

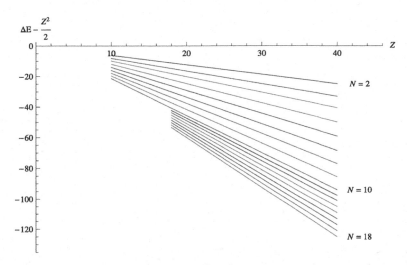

Fig. 4.4 For isoelectronic series, equation (4.60) indicates that within the large-Z approximation, the quantity $\Delta E - Z^2/2$ is exactly linear in Z, as is illustrated above. ΔE is the core ionization energy.

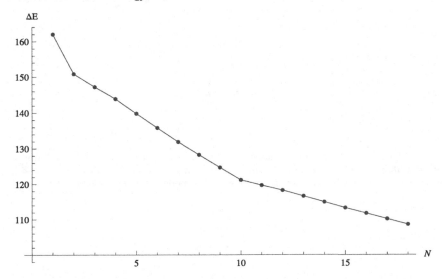

Fig. 4.5 For isonuclear series, the dependence of the core ionization energy on N is approximately piecewise linear. Whenever a new shell starts to fill, the slope of the line changes. The dots in the figure were calculated using equation (4.60), where it is not obvious that the dependence ought to be approximately piecewise linear. However, equations (4.57) and (4.55) can give us some insight into the approximately piecewise linear relationship.

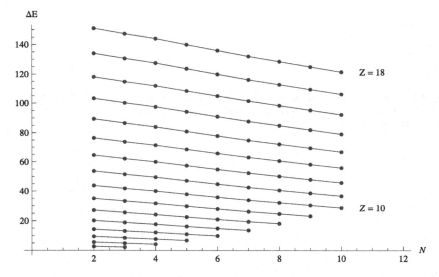

Fig. 4.6   This figure shows the dependence of the core-ionization energy on both N and Z for the filling of the $n = 2$ shell. Points with $N > Z$ are omitted because the large-Z approximation cannot be used for these points. The energies are in Hartrees.

4.6   Advantages and disadvantages of Goscinskian configurations

We seen that when $V_0(\mathbf{x})$ is chosen to be the Coulomb attraction of the bare nucleus, the approximate Schrödinger equation

$$\left[-\frac{1}{2}\sum_{j=1}^{N}\nabla_j^2 + \beta_\nu V_0(\mathbf{x}) - E_\kappa\right]|\Phi_\nu\rangle = 0 \qquad \beta_\nu V_0(\mathbf{x}) = -\sum_{j=1}^{N}\frac{Q_\nu}{r_j} \quad (4.61)$$

can be solved exactly using configurations composed of hydrogenlike spin-orbitals with the especially chosen weighted charges Q_ν shown in equation (4.4). There is no need to calculate the weighting factors β_ν. These are obtained automatically when the secular equation is solved. Nor is there a need to normalize the configurations. This is also achieved automatically. Thus the choice of $V_0(\mathbf{x})$ as the potential of the bare nucleus has many advantages; but it also has disadvantages. Just as is the case in perturbation theory, convergence is most rapid if $V_0(\mathbf{x})$ is chosen to be as close as possible to the actual potential. By choosing $V_0(\mathbf{x})$ to be the Coulomb attraction of the nucleus, we have neglected interelectron repulsion. This is why the Generalized Sturmian Method with Goscinskian configurations works best when the number of electrons in an atom or ion is small, and why it works especially well when $Z >> N$, i.e. when the Coulomb attraction of the nucleus dominates over the effects of interelectron repulsion.

To extend the range of applicability of the method to atoms and ions with large values of N, we would need to choose a $V_0(\mathbf{x})$ which included some of the effects of interelectron repulsion. For example, we could let it be the Hartree potential. The approximate Schrödinger equation (4.6) can always be solved provided that it is separable, and it is separable whenever the approximate potential has the form

$$V_0(\mathbf{x}) = \sum_{j=1}^{N} v(\mathbf{x}_j) \qquad (4.62)$$

The separated form of (4.6) becomes:

$$\left[-\frac{1}{2}\nabla_j^2 + \beta_\nu v(\mathbf{x}_j) - \epsilon_\zeta\right]\varphi_\zeta(\mathbf{x}_j) = 0 \qquad (4.63)$$

where the weighting factors β_ν must be chosen in such a way that

$$\sum_{\zeta \in \nu} \epsilon_\zeta = E_\kappa \qquad (4.64)$$

If the spin-orbitals $\varphi_\zeta(\mathbf{x}_j)$ satisfy (4.63), then configurations of the form

$$|\Phi_\nu\rangle = |\varphi_\zeta \varphi_{\zeta'} \varphi_{\zeta''} \cdots | \equiv \frac{1}{\sqrt{N!}} \begin{vmatrix} \varphi_\zeta(\mathbf{x}_1) & \varphi_{\zeta'}(\mathbf{x}_1) & \varphi_{\zeta''}(\mathbf{x}_1) & \cdots \\ \varphi_\zeta(\mathbf{x}_2) & \varphi_{\zeta'}(\mathbf{x}_2) & \varphi_{\zeta''}(\mathbf{x}_2) & \cdots \\ \varphi_\zeta(\mathbf{x}_3) & \varphi_{\zeta'}(\mathbf{x}_3) & \varphi_{\zeta''}(\mathbf{x}_3) & \cdots \\ \vdots & \vdots & \vdots & \end{vmatrix} \qquad (4.65)$$

will satisfy the approximate Schrödinger equation (4.6). Some of the neatness of the Generalized Sturmian Method with Goscinskian configurations is certainly lost by choosing a $V_0(\mathbf{x})$ that includes effects of interelectron repulsion, but it could be worth paying this price in order to extend the method to atoms and atomic ions with larger values of N. We are at present exploring these possibilities, and some work in this direction is also being done by Prof. Gustavo Gasaneo and his group in Argentina.

4.7 $\mathcal{R}$-blocks, invariant subsets and invariant blocks

To tie the discussion of this chapter in with the general principles discussed in Chapter 1, we identify T with the operator whose roots and eigenfunctions we wish to study. The group of symmetry operations $\mathcal{G}$ that leave the nuclear attraction and interelectron repulsion matrix of an atom invariant consists of rotations of the entire system about the nucleus, together with reflections and inversions that do not affect the interelectron distances. These operations do not affect the radial parts of the atomic orbitals from which the Goscinskian configurations are constructed, nor do they affect the spin. Thus the set of configurations, all of which are characterized by the same value of

$$\mathcal{R}_\nu \equiv \sqrt{\frac{1}{n^2} + \frac{1}{n'^2} + \frac{1}{n''^2} + \cdots} \qquad (4.66)$$

i.e. configurations all of which are built from hydrogenlike atomic spin-orbitals with a particular set of principal quantum numbers $(n, n', n'', \ldots)$, is closed under $\mathcal{G}$, and it corresponds to an invariant subset as discussed in Chapter 1. The block of T' based on it corresponds to an invariant block. As expected, the eigenfunctions of interelectron repulsion matrix for the $\mathcal{R}$-blocks are the symmetry-adapted basis functions that we desire. In Chapter 1, we mentioned that when the roots of an invariant block are degenerate, then in order to take full advantage of the symmetry of the problem, we need to add an extremely small perturbation which will slightly remove the degeneracy. In the present case, this slight perturbation is given by

$$T_p = aL_z + bS_z F \qquad (4.67)$$

where a and b are two very small irrational numbers. (They are chosen to be irrational in order to avoid accidental degeneracies.) When this small perturbation is added to T', the degeneracy is slightly removed. The eigenfunctions of $T' + T_p$ for an $\mathcal{R}$-block are then Russell-Saunders states, i.e. they are simultaneous eigenfunctions of the total angular momentum operator L^2, its z-component L_z, the total spin operator S^2, and its z-component S_z. We can ask how many linearly independent configurations there are in a ground-state $\mathcal{R}$-block. The answer is that when the Pauli principle is taken into account, the number of configurations m_k in an $\mathcal{R}$-block is given by the binomial coefficient

$$\binom{n_s}{N_v} = \frac{n_s!}{N_v!(n_s - N_v)!} \equiv m_k \qquad (4.68)$$

where n_s is the number of atomic spin-orbitals in the highest filled shell, while N_v is the number of valence electrons. For the ground-state configurations illustrated in Tables 4.3 and 4.4, N_v is the number of valence electrons. For example, for the ground state of lithium, we are starting to fill the $n = 2$ shell, for which $n_s=8$. Since there is only one valence electron, $N_v=1$ and

$$m_k \text{ for lithium ground state:} \quad \binom{n_s}{N_v} = \binom{8}{1} = \frac{8!}{1!(8-1)!} = 8 \quad (4.69)$$

Tables 4.3 and 4.4 show the roots of T' and their corresponding spectral terms for the first row of the periodic table. For the lithiumlike isoelectronic series, one of the roots corresponds to a ^{2}S state and the other corresponds to a ^{2}P state. We now remember that the total degeneracy of a Russell-Saunders state is given by $(2L+1) \times (2S+1)$, where L and S are respectively the quantum numbers of total orbital angular momentum and total spin. The ^2S state is thus 2-fold degenerate, while the ^2P state is 6-fold degenerate. By diagonalizing the ground-state $\mathcal{R}$-block for lithium, we thus obtain 8 Russell-Saunders states, the same as the number of Pauli-allowed states in the block. With a little effort, the reader can verify that the number of Russell-Saunders states shown in Tables 4.3 and 4.4 corresponds to the number of Pauli-allowed configurations given by the binomial coefficient in equation (4.68). Of course for the lithium ground state, the construction of Russell-Saunders states is trivial, while for beryllium it can be accomplished with the help of Clebsch-Gordan coefficients. However for carbon,

Table 4.2: Eigenvalues of the 2-electron interelectron repulsion matrix $T'_{\nu',\nu}$ for $S=1$, $M_S=1$, $n = 2$ and $n'=3, 4, 5$.

$n'=3$ $\|\lambda_\kappa\|$	term	$n' = 4$ $\|\lambda_\kappa\|$	term	$n'=5$ $\|\lambda_\kappa\|$	term
.108252	^{3}S	.077484	^{3}S	.056075	^{3}S
.134734	^{3}P	.087582	^{3}P	.065019	^{3}P
.135408	^{3}D	.090845	^{3}D	.061128	^{3}P
.138421	^{3}P	.093401	^{3}P	.063370	^{3}D
.155155	^{3}F	.099235	^{3}F	.067758	^{3}F
.160439	^{3}P	.099991	^{3}P	.067934	^{3}P
.165613	^{3}D	.104253	^{3}D	.070494	^{3}D
.168814	^{3}S	.106271	^{3}D	.071269	^{3}D
.173917	^{3}D	.107976	^{3}S	.072413	^{3}F
.186893	^{3}P	.108188	^{3}F	.072857	^{3}S
		.111210	^{3}G	.073295	^{3}G
		.111264	^{3}F	.073588	^{3}G
		.113313	^{3}P	.073920	^{3}F
		.114381	^{3}D	.074306	^{3}G
				.074578	^{3}H
				.074963	^{3}F
				.075173	^{3}P
				.075545	^{3}D

where

$$m_k \text{ for carbon ground state: } \binom{n_s}{N_v} = \binom{8}{4} = \frac{8!}{4!(8-4)!} = 70 \quad (4.70)$$

the construction of the Russell-Saunders states is non-trivial. An indication of the nature of the states obtained is given in Table 4.5. As can easily be verified, the sum of the degeneracies shown in the table is equal to 70.

Table 4.3: Roots of the ground state $\mathcal{R}$-block of the interelectron repulsion matrix for the Li-like, Be-like, B-like and C-like isoelectronic series.

Li-like $|\lambda_\kappa|$	term	Be-like $|\lambda_\kappa|$	term	B-like $|\lambda_\kappa|$	term	C-like $|\lambda_\kappa|$	term
0.681870	^{2}S	0.986172	^{1}S	1.40355	^{2}P	1.88151	^{3}P
0.729017	^{2}P	1.02720	^{3}P	1.44095	^{4}P	1.89369	^{1}D
		1.06426	^{1}P	1.47134	^{2}D	1.90681	^{1}S
		1.09169	^{3}P	1.49042	^{2}S	1.91623	^{5}S
		1.10503	^{1}D	1.49395	^{2}P	1.995141	^{3}D
		1.13246	^{1}S	1.52129	^{4}S	1.96359	^{3}P
				1.54037	^{2}D	1.98389	^{3}S
				1.55726	^{2}P	1.98524	^{1}D
						1.99742	^{1}P
						2.04342	^{3}P
						2.05560	^{1}D
						2.07900	^{1}S

Table 4.4: Roots of the ground state $\mathcal{R}$-block of the interelectron repulsion matrix $T'_{\nu'\nu}$ for the N-like, O-like, F-like and Ne-like isoelectronic series.

N-like $\|\lambda_\kappa\|$	term	O-like $\|\lambda_\kappa\|$	term	F-like $\|\lambda_\kappa\|$	term	Ne-like $\|\lambda_\kappa\|$	term
2.41491	^{4}S	3.02641	^{3}P	3.68415	^{2}P	4.38541	^{1}S
2.43246	^{2}D	3.03769	^{1}D	3.78926	^{2}S		
2.44111	^{2}P	3.05065	^{1}S				
2.49314	^{4}P	3.11850	^{3}P				
2.52109	^{2}D	3.14982	^{1}P				
2.53864	^{2}S	3.24065	^{1}S				
2.54189	^{2}P						
2.61775	^{2}P						

Table 4.5: Eigenvalues of $T'_{\nu',\nu}$ for the carbon-like $\mathcal{R}_\nu = \sqrt{3}$ block.

$|\lambda_\kappa|$	term	degen.	configuration
1.88151	^3P	9	$.994467(1s)^2(2s)^2(2p)^2+.105047(1s)^2(2p)^4$
1.89369	^1D	5	$.994467(1s)^2(2s)^2(2p)^2-.105047(1s)^2(2p)^4$
1.90681	^1S	1	$.979686(1s)^2(2s)^2(2p)^2+.200537(1s)^2(2p)^4$
1.91623	^5S	5	$(1s)^2(2s)(2p)^3$
1.95141	^3D	15	$(1s)^2(2s)(2p)^3$
1.96359	^3P	9	$(1s)^2(2s)(2p)^3$
1.98389	^3S	3	$(1s)^2(2s)(2p)^3$
1.98524	^1D	5	$(1s)^2(2s)(2p)^3$
1.99742	^1P	3	$(1s)^2(2s)(2p)^3$
2.04342	^3P	9	$.105047(1s)^2(2s)^2(2p)^2-.994467(1s)^2(2p)^4$
2.05560	^1D	5	$.105047(1s)^2(2s)^2(2p)^2+.994467(1s)^2(2p)^4$
2.07900	^1S	1	$.200537(1s)^2(2s)^2(2p)^2-.979686(1s)^2(2p)^4$

4.8 Invariant subsets based on subshells; Classification according to M_L and M_s

When we are dealing with configurations corresponding to excited states of an atom, the number of Pauli-allowed states in an $\mathcal{R}$-block may be very large, and we can ask whether our large basis set of Goscinskian configurations spanning W can be divided into smaller invariant subsets spanning the subspaces W_k. A little thought reveals that this is indeed possible. The symmetry operations that leave T' invariant do not affect the radial parts $R_{n,l}(r)$ of the atomic orbitals, so sets of configurations built from the subshell sets $((n, l), (n'l'), (n'', l''), \ldots)$ will be invariant subsets of our large basis set W under the operations of $\mathcal{G}$. The results will then differ slightly from those obtained in the large-Z approximation, which allows mixing between subshells. As an example of an invariant subset based on subshells we can consider the set of configurations corresponding to $(1s)(2d)^2$ for the lithium-like isoelectronic series:

$$(1s)(3d)^2 \qquad \binom{2}{1} \times \binom{10}{2} = 90 \qquad (4.71)$$

The invariant subset contains 90 configurations. Diagonalization of the 90×90 block yields the Russell Saunders states shown in Table 4.5.

We can also pick subspaces W_k characterized by particular eigenvalues of S_z and L_z. These reductions in the size of the invariant subsets and the invariant blocks make it feasible to generate symmetry-adapted basis sets automatically also in the case of highly excited configurations. The use of symmetry-adapted basis sets leads to accurate calculations as is illustrated in Tables 4.7-4.9.

Table 4.6: This table shows the multiplets generated by diagonalizing the energy-invariant interelectron repulsion matrix T' for the 90×90 block the Hamiltonian based on neutral lithium configurations corresponding to $(1s)(3d)^2$, with $R_\nu = \sqrt{11}/3$. The reader can verify that the sum of the degeneracies of the multiplets is 90. The energies shown are for neutral lithium.

$|\lambda_\kappa|$	term	degen.	energy
.270978	^{4}F	28	−4.63798
.271649	^{2}F	14	−4.63594
.278128	^{2}D	10	−4.63526
.278998	^{4}P	12	−4.61623
.279669	^{2}P	6	−4.61359
.281871	^{2}G	18	−4.61155
.297850	^{2}S	2	−4.55650

Table 4.7: ^3S excited state energies calculated with 78 Goscinskians, using the crude relativistic correction described in the text. The calculation of similar tables for ^1P, ^3P, ^1D, ^3D, doubly excited autoionizing states, etc., is equally easy, rapid, and of comparable accuracy. Tables are given in Chapter 4 in [Avery and Avery, 2006], but may easily be reproduced using our programs, as shown in Tutorial 1 on [Avery and Avery, 2006a].

	He	Li$^+$	Be^{2+}	B^{3+}	C^{4+}	N^{5+}
1s2s ^3S	-2.1737	-5.1085	-9.2957	-14.735	-21.427	-29.373
expt.	-2.1750	-5.1109	-9.2983	-14.738	-21.429	-29.375
1s3s ^3S	-2.0683	-4.7509	-8.5459	-13.454	-19.476	-26.612
expt.	-2.0685	-4.7522	-8.5480	-13.457	-19.478	-26.614
1s4s ^3S	-2.0364	-4.6365	-8.2999	-13.027	-18.820	-25.678
expt.	-2.0363	-4.6373	-8.3015	-13.030	-18.822	-25.680
1s5s ^3S	-2.0226	-4.5859	-8.1896	-12.835	-18.522	-25.253
expt.	-2.0224	-4.5862	-8.1905		-18.524	-25.254
1s6s ^3S	-2.0154	-4.5591	-8.1309	-12.732	-18.363	-25.024
expt.	-2.0152	-4.5592			-18.364	
1s7s ^3S	-2.0112	-4.5432	-8.096	-12.67	-18.267	-24.888
expt.	-2.0109	-4.5431			-18.268	
1s8s ^3S	-2.0085	-4.5330	-8.0736	-12.631	-18.206	-24.799
expt.	-2.0082	-4.5328			-18.206	
1s9s ^3S	-2.0067	-4.5261	-8.0583	-12.604	-18.164	-24.739
expt.	-2.0064					
1s10s ^3S	-2.0051	-4.5212	-8.0475	-12.585	-18.134	-24.696
expt.	-2.0051					

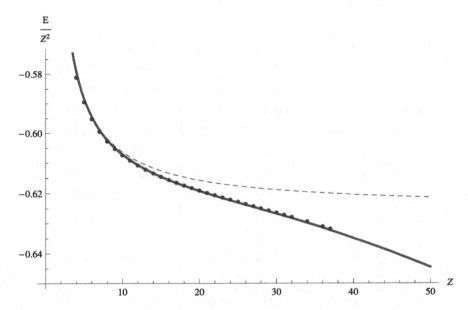

Fig. 4.7   Here the lowest ^{3}S energies of the heliumlike isoelectronic series are divided by Z^2. The lower line is corrected for relativistic effects. The dots are experimental values.

Table 4.8: ^1S ground and excited state energies (in Hartrees) for the 2-electron iso-electronic series. The basis set used consisted of 592 generalized Sturmians of the Goscinski type, using the crude relativistic correction described in the text. The whole table was computed approximately a second. Experimental values are taken from the NIST tables [National Institute for Standards and Technology (NIST)] (http://physics.nist.gov/asd).

	He	Li$^+$	Be^{2+}	B^{3+}	C^{4+}
1s^2 ^{1}S	−2.8956	−7.2716	−13.649	−22.028	−32.412
expt.	−2.9034	−7.2798	−13.657	−22.035	−32.416
1s2s ^{1}S	−2.1441	−5.0350	−9.1768	−14.571	−21.218
expt.	−2.1458	−5.0410	−9.1860	−14.582	−21.230
1s3s ^{1}S	−2.0607	−4.7303	−8.5112	−13.405	−19.414
expt.	−2.0611	−4.7339	−8.5183	−13.415	−19.425
1s4s ^{1}S	−2.0333	−4.6280	−8.2844	−13.005	−18.791
expt.	−2.0334	−4.6299	−8.2891		−18.800
1s5s ^{1}S	−2.0221	−4.5858	−8.1889	−12.833	−18.520
expt.	−2.0210	−4.5825			−18.513
1s6s ^{1}S	−2.0147	−4.5579	−8.129	−12.729	−18.359
expt.	−2.0144	−4.5571			
1s7s ^{3}S	−2.0109	−4.5426	−8.0951	−12.67	−18.281
expt.	−2.0104	−4.5418			
1s8s ^{3}S	−2.0083	−4.5326	−8.0732	−12.641	−18.26
expt.	−2.0079				
1s9s ^{3}S	−2.0065	−4.5258	−8.0583	−12.626	−18.203
expt.	−2.0062				
1s10s ^{3}S	−2.0049	−4.521	−8.0476	−12.602	−18.162
expt.	−2.0050				

Table 4.9: ^1D excited state energies for the 2-electron isoelectronic series, compared with experimental values taken from the NIST tables.

	He	Li$^+$	Be^{2+}	B^{3+}	C^{4+}	N^{5+}
1s3d ^{1}D	−2.0555	−4.7218	−8.4990	−13.388	−19.387	−26.498
expt.	−2.0554	−4.7225	−8.5012	−13.392	−19.396	−26.514
1s4d ^{1}D	−2.0312	−4.6246	−8.2801	−12.998	−18.779	−25.622
expt.	−2.0311	−4.6252	−8.2824	−13.003	−18.788	−25.639
1s5d ^{1}D	−2.0200	−4.5797	−8.1790	−12.818	−18.497	−25.217
expt.	−2.0198	−4.5801	−8.1807		−18.507	−25.234
1s6d ^{1}D	−2.0139	−4.5554	−8.1242	−12.721	−18.345	−24.997
expt.	−2.0137	−4.5557			−18.354	
1s7d ^{1}D	−2.0102	−4.5407	−8.0912	−12.662	−18.253	−24.865
expt.	−2.0100	−4.5409			−18.262	
1s8d ^{1}D	−2.0078	−4.5311	−8.0699	−12.624	−18.194	−24.779
expt.	−2.0076	−4.5314			−18.202	
1s9d ^{1}D	−2.0062	−4.5246	−8.0552	−12.598	−18.153	−24.720
expt.	−2.0060					
1s10d ^{1}D	−2.0050	−4.5199	−8.0447	−12.579	−18.124	−24.678
expt.	−2.0048					
1s11d ^{1}D	−2.0041	−4.5165	−8.0370	−12.566	−18.102	−24.647
expt.	−2.0035					
1s12d ^{1}D	−2.0032	−4.5139	−8.0311	−12.555	−18.086	−24.624
expt.	−2.0033					

4.9   An atom surrounded by point charges

For a heavy atom surrounded by lighter atoms (ligands), the effect of the ligands is sometimes approximated by considering the atom to be surrounded by an array of point charges with an appropriate symmetry. The potential experienced by such an atom is

$$V(\mathbf{x}) = V_0(\mathbf{x}) + V'(\mathbf{x}) + V''(\mathbf{x}) \tag{4.72}$$

where V_0 is the nuclear attraction potential

$$V_0(\mathbf{x}) = -\sum_{i=1}^{N} \frac{Z_0}{r_i} \tag{4.73}$$

and V' is the interelectron repulsion potential

$$V'(\mathbf{x}) = \sum_{j>i}^{N} \sum_{i=1}^{N} \frac{1}{r_{ij}} \tag{4.74}$$

while V'' (the "crystal field") expresses the effect of a set of charges q_a located at the positions $\mathbf{X}_a$:

$$V''(\mathbf{x}) = -\sum_{i=1}^{N} \sum_{a} \frac{q_a}{|\mathbf{x}_i - \mathbf{X}_a|} \tag{4.75}$$

The generalized Sturmian secular equation analogous to (4.34) then becomes

$$\sum_{\nu} \left[\delta_{\nu',\nu} Z\mathcal{R}_\nu + T'_{\nu',\nu} + T''_{\nu',\nu} - p_\kappa \delta_{\nu',\nu} \right] C_{\nu,\kappa} = 0 \tag{4.76}$$

where

$$T''_{\nu',\nu} \equiv -\frac{1}{p_\kappa} \langle \Phi^*_{\nu'} | V'' | \Phi_\nu \rangle \tag{4.77}$$

In order to evaluate $T''_{\nu',\nu}$ we must first calculate one-electron matrix elements of the form

$$
\begin{aligned}
v''_{\mu_1,\mu_2} &\equiv \int d^3x \; \chi^*_{\mu_1}(\mathbf{x}) \chi_{\mu_2}(\mathbf{x}) \sum_a \frac{q_a}{|\mathbf{x} - \mathbf{X}_a|} \\
&= \sum_a \sum_l q_a \int d^3x_i \; \chi^*_{\mu_1}(\mathbf{x}) \chi_{\mu_2}(\mathbf{x}) \left(\frac{r_<^l}{r_>^{l+1}} \right)_a P_l(\hat{\mathbf{x}} \cdot \hat{\mathbf{X}}_a) \\
&= \sum_a \sum_l q_a \int_0^\infty dr \; r^2 \left(\frac{r_<^l}{r_>^{l+1}} \right)_a R_{n_1,l_1}(r) R_{n_2,l_2}(r) \\
&\quad \times \int d\Omega \; Y^*_{l_1,m_1}(\hat{\mathbf{x}}) Y_{l_2,m_2}(\hat{\mathbf{x}}) P_l(\hat{\mathbf{x}} \cdot \hat{\mathbf{X}}_a)
\end{aligned}
\tag{4.78}
$$

where

$$\left(\frac{r_<^l}{r_>^{l+1}}\right)_a \equiv \begin{cases} r^l/R_a^{l+1} & r < R_a \\ \\ R_a^l/r^{l+1} & R_a < r \end{cases} \tag{4.79}$$

If the points $\mathbf{X}_a$ are all equidistant from the central atom and if the charges are all equal, then equation (4.78) can be rewritten in the form:

$$
\begin{aligned}
v''_{\mu_1,\mu_2} &= \sum_a \sum_l \frac{4\pi q_a}{2l+1} \int_0^\infty dr \; r^2 \left(\frac{r_<^l}{r_>^{l+1}}\right)_a R_{n_1,l_1}(r)R_{n_2,l_2}(r) \\
&\quad \times Y_{l,m}^*(\hat{\mathbf{X}}_a) \int d\Omega \; Y_{l_1,m_1}^*(\hat{\mathbf{x}})Y_{l_2,m_2}(\hat{\mathbf{x}})Y_{l,m}(\hat{\mathbf{x}}) \\
&\equiv \sum_l F_l \sum_a Y_{l,m}^*(\hat{\mathbf{X}}_a) \int d\Omega \; Y_{l_1,m_1}^*(\hat{\mathbf{x}})Y_{l_2,m_2}(\hat{\mathbf{x}})Y_{l,m}(\hat{\mathbf{x}})
\end{aligned}
\tag{4.80}
$$

where

$$F_l \equiv \frac{4\pi q_a}{2l+1} \int_0^\infty dr \; r^2 \left(\frac{r_<^l}{r_>^{l+1}}\right)_a R_{n_1,l_1}(r)R_{n_2,l_2}(r) \tag{4.81}$$

Once we are in possession of v''_{μ_1,μ_2}, the matrix element T''_{ν_1,ν_2} can be evaluated by means of the generalized Slater-Condon rules discussed in Appendix D, Section D.1. For fixed angular geometry, the matrix T''_{ν_1,ν_2} turns out to be a function of the parameters

$$s_a \equiv p_\kappa R_a \equiv p_\kappa |\mathbf{X}_a| \tag{4.82}$$

where $|\mathbf{X}_a|$ is the distance of the ath charge q_a from the central atom. The generalized Sturmian secular equations (4.76) can be solved as follows: We begin by picking values of s_a The secular equation (4.76) is then solved, yielding roots p_κ for the ground state and the excited states. The corresponding values of $|\mathbf{X}_a|$ are then known. This can be repeated for a number of s_a values, yielding solutions as functions of the distances $|\mathbf{X}_a|$. As an example, we can consider an atom surrounded by 8 equal point charges q at the positions

$$
\begin{aligned}
\hat{\mathbf{X}}_1 &= (+1,+1,+1)/\sqrt{3} & \hat{\mathbf{X}}_5 &= (+1,-1,-1)/\sqrt{3} \\
\hat{\mathbf{X}}_2 &= (-1,+1,+1)/\sqrt{3} & \hat{\mathbf{X}}_6 &= (-1,+1,-1)/\sqrt{3} \\
\hat{\mathbf{X}}_3 &= (+1,-1,+1)/\sqrt{3} & \hat{\mathbf{X}}_7 &= (-1,-1,+1)/\sqrt{3} \\
\hat{\mathbf{X}}_4 &= (+1,+1,-1)/\sqrt{3} & \hat{\mathbf{X}}_8 &= (-1,-1,-1)/\sqrt{3}
\end{aligned}
\tag{4.83}
$$

which has cubic symmetry. We can also discuss the effect of a square-planar array

$$\hat{\mathbf{X}}_1 = (+1, 0, 0) \qquad \hat{\mathbf{X}}_3 = (0, +1, 0)$$
$$\hat{\mathbf{X}}_2 = (-1, 0, 0) \qquad \hat{\mathbf{X}}_4 = (0, -1, 0) \tag{4.84}$$

a linear array:

$$\hat{\mathbf{X}}_1 = (0, 0, +1) \qquad \hat{\mathbf{X}}_2 = (0, 0, -1) \tag{4.85}$$

a trigonal array:

$$\hat{\mathbf{X}}_1 = (2\sqrt{2}/3, 0, -1/3)$$
$$\hat{\mathbf{X}}_2 = (-\sqrt{2}/3, \sqrt{6}/3, -1/3)$$
$$\hat{\mathbf{X}}_3 = (-\sqrt{2}/3, -\sqrt{6}/3, -1/3) \tag{4.86}$$

and a tetrahedral array

$$\hat{\mathbf{X}}_1 = (0, 0, 1)$$
$$\hat{\mathbf{X}}_2 = (2\sqrt{2}/3, 0, -1/3)$$
$$\hat{\mathbf{X}}_3 = (-\sqrt{2}/3, \sqrt{2/3}, -1/3)$$
$$\hat{\mathbf{X}}_4 = (-\sqrt{2}/3, -\sqrt{2/3}, -1/3) \tag{4.87}$$

The presence of a symmetrical set of charges surrounding a central atom reduces the symmetry of the system: Instead of being invariant under the full group of rotations about the central atom (as well as rotations in spin space) the system is now invariant only under a restricted group of rotations. However, subsets of configurations that are closed under the operations of the full rotation group, are also closed under the more restricted group of rotations. Therefore, if we are willing to have invariant subsets of configurations that are larger than minimal, we can choose them almost as though the perturbing surrounding charges were not there. Thus the invariant subsets in the perturbed case can be established by following the prescription:

$$\nu = (a; b)$$
$$a = (M_S, (n, l), (n', l'), (n'', l''), \ldots)$$
$$b = ((m, m_s), (m', m'_s), (m'', m''_s), \ldots) \tag{4.88}$$
$$W_a = \text{span}\left\{ |\Phi_{a,b_1}\rangle, |\Phi_{a,b_2}\rangle, |\Phi_{a,b_3}\rangle, \ldots \right\}$$

The symmetry-adapted basis functions found by diagonalizing the invariant blocks can be used as basis functions for a more ambitious configuration interaction calculation. For example, let us think of a neutral lithium atom, surrounded by a square planar array of point charges. To make the example more specific, let us think of the set of configurations characterized by

$$(1s)^2(3d) \qquad \binom{10}{1} \qquad (4.89)$$

This set has 10 members, and they can be used as a basis for solving the generalized Sturmian secular equations in the presence of the square-planar array of point charges. In the absence of the charges, the 10 configurations are members of a ^2D multiplet, and in the large-Z approximation its energy is

$$E_\kappa = -\frac{1}{2}(ZR_\nu - |\lambda|)^2 = -\frac{1}{2}(\sqrt{19} - .58281)^2 = -7.1294 \qquad (4.90)$$

This corresponds to the effective charge

$$Q = \frac{p_\kappa}{R_\nu} = \frac{ZR_\nu - |\lambda|}{R_\nu} = 2.5989 \qquad (4.91)$$

where

$$R_\nu = \sqrt{\frac{1}{1^2} + \frac{1}{1^2} + \frac{1}{3^2}} = \frac{\sqrt{19}}{3} \qquad |\lambda| = .58281 \qquad (4.92)$$

In order to calculate T'' and find the effect of the point charges (which we treat here as a small perturbation), we need to evaluate the radial integrals:

$$\int_0^\infty dr\, r^2 \left(\frac{r_<^l}{r_>^{l+1}}\right)_a |R_{3,2}(r)|^2 = \begin{cases} .19193 & l=0 \\ .10363 & l=2 \\ .06826 & l=4 \end{cases} \qquad (4.93)$$

with $|\mathbf{X}_a| = 5$ and

$$R_{3,2}(r) = \frac{2Q^{7/2}}{81}\sqrt{\frac{2}{15}}\, r^2 e^{-Qr/3} \qquad (4.94)$$

The angular integrals needed for constructing T'' are

$$\sum_a \int d\Omega\, Y_{2,m_1}^*(\hat{\mathbf{x}})Y_{2,m_2}(\hat{\mathbf{x}})P_0(\hat{\mathbf{x}}\cdot\hat{\mathbf{X}}_a) = 4\delta_{m_1,m_2}$$

$$\sum_a \int d\Omega\, Y_{2,m_1}^*(\hat{\mathbf{x}})Y_{2,m_2}(\hat{\mathbf{x}})P_1(\hat{\mathbf{x}}\cdot\hat{\mathbf{X}}_a) = 0 \qquad (4.95)$$

$$\sum_a \int d\Omega\, Y_{2,m_1}^*(\hat{\mathbf{x}})Y_{2,m_2}(\hat{\mathbf{x}})P_3(\hat{\mathbf{x}}\cdot\hat{\mathbf{X}}_a) = 0$$

and

$$\sum_a \int d\Omega \; Y_{2,m_1}^*(\hat{\mathbf{x}}) Y_{2,m_2}(\hat{\mathbf{x}}) P_2(\hat{\mathbf{x}} \cdot \hat{\mathbf{X}}_a)$$

$$= \begin{pmatrix} \dfrac{4}{7} & 0 & 0 & 0 & 0 \\[2ex] 0 & -\dfrac{2}{7} & 0 & 0 & 0 \\[2ex] 0 & 0 & -\dfrac{4}{7} & 0 & 0 \\[2ex] 0 & 0 & 0 & -\dfrac{2}{7} & 0 \\[2ex] 0 & 0 & 0 & 0 & \dfrac{4}{7} \end{pmatrix} \tag{4.96}$$

while

$$\sum_a \int d\Omega \; Y_{2,m_1}^*(\hat{\mathbf{x}}) Y_{2,m_2}(\hat{\mathbf{x}}) P_4(\hat{\mathbf{x}} \cdot \hat{\mathbf{X}}_a)$$

$$= \begin{pmatrix} \dfrac{1}{14} & 0 & 0 & 0 & \dfrac{5}{6} \\[2ex] 0 & -\dfrac{2}{7} & 0 & 0 & 0 \\[2ex] 0 & 0 & \dfrac{3}{7} & 0 & 0 \\[2ex] 0 & 0 & 0 & -\dfrac{2}{7} & 0 \\[2ex] \dfrac{5}{6} & 0 & 0 & 0 & \dfrac{1}{14} \end{pmatrix} \tag{4.97}$$

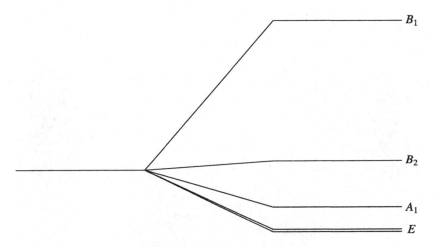

Fig. 4.8   *This diagram shows schematically the splitting of the $^2D_{1/2}$ multiplet of neutral lithium discussed above in the field of a square-planar array of point charges. The $^2D_{-1/2}$ multiplet which is similarly split is not shown. Besides being split by the presence of the charges, the multiplet is also displaced in energy, as is the lithium ground state.*

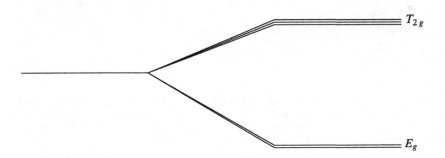

Fig. 4.9   *This figure shows the same multiplet as Figure 4.8, but the splitting is due to an octohedral array of point charges.*

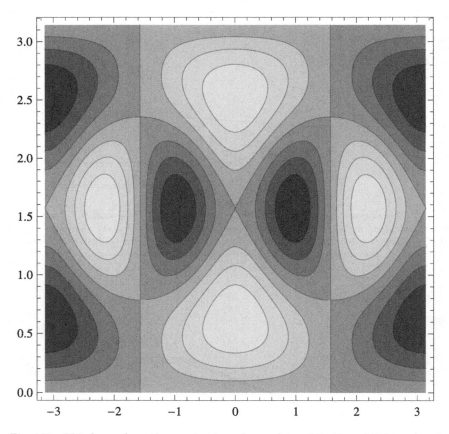

Fig. 4.10   This figure shows the angular dependence of one of the T_{1u} orbitals produced when degenerate f-orbitals are split by the presence of an octohedral arrangement of point charges.   The T_{1u} irreducible representation of the group O_h is a 3-dimensional representation.   The other two components are shown in Figure 4.11.

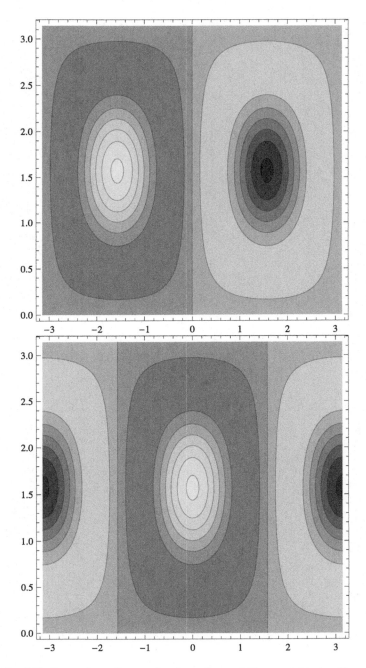

Fig. 4.11 *This figure shows the angular dependence of the remaining two T_{1u} orbitals. Both in this figure and in Figure 4.10, the vertical axis represents θ and the horizontal axis represents ϕ.*

Chapter 5

MOLECULAR ORBITALS BASED ON STURMIANS

5.1 The one-electron secular equation

Molecular orbitals may be represented as superpositions of Coulomb Sturmian basis functions (Appendix B) centered on the nuclei of a molecule. These basis functions are an example of Exponential-Type Orbitals (ETO's) [Harris and Michels, 1967], [Pinchon and Hoggan 2007], [Pinchon and Hoggan, 2009], [Weatherford, 1982], and calculations using them can potentially be much more accurate than calculations based on Gaussians.

Gaussian basis functions have serious drawbacks, since very many of them are needed to approximate the molecular orbitals, and since the cusp at the nucleus is never adequately represented. Furthermore, Gaussian basis functions cannot accurately represent the exponential decay of the orbitals at large distances from the nuclei. Thus while the mainstream effort of quantum chemistry today follows the path of Gaussian technology, there exists a small group of researchers who struggle with the difficult mathematical problems involved in using exponential-type orbitals (ETO's) as basis functions, and we hope that the present chapter will make a contribution to this effort. We will see that the automatic scaling properties associated with the Generalized Sturmian Method have advantages also in the case of molecules, and that molecular orbitals based on many-center Coulomb Sturmians have advantages over other ETO's with respect to the ease of evaluation of interelectron repulsion integrals.

We will first consider the use of Coulomb Sturmian basis functions located on the different atoms of a molecule to solve the 1-electron molecular Schrödinger equation, an endeavor which was pioneered by C.E, Wulfman, B. Judd, T. Koga, V. Aquilanti and others [Shibuya and Wulfman, 1965], [Wulfman, 2011], [Judd, 1975], [Koga et al., 1984]-[Koga et al., 1991],

[Aquilanti et al., 1996],[Aquilanti et al., 1997]. These authors solved the Schrödinger equation in momentum space, but here we will use a direct-space treatment to reach the same results. In this approach to molecular orbital theory, we search for solutions to the one-electron Schrödinger equation

$$\left[-\frac{1}{2}\nabla^2 + v(\mathbf{x}) - \epsilon_\zeta \right] \varphi_\zeta(\mathbf{x}) = 0 \qquad (5.1)$$

where $v(\mathbf{x})$ is the Coulomb attraction potential of the nuclei:

$$v(\mathbf{x}) = -\sum_a \frac{Z_a}{|\mathbf{x} - \mathbf{X}_a|} \qquad (5.2)$$

We will approximate the molecular orbitals $\varphi_\zeta(\mathbf{x})$ by superpositions of Coulomb Sturmian atomic orbitals centered on the various atoms of the molecule. To do this it is convenient to introduce a notation where τ stands for a set of four indices, the first three being the quantum numbers of a one-electron Coulomb Sturmian basis function of the type discussed in Appendix B, while the final index, a, is the index of the nucleus on which the atomic orbital is centered:

$$\tau \equiv (n, l, m, a) \qquad (5.3)$$

In this notation we can write

$$\chi_\tau(\mathbf{x}) \equiv \chi_{nlm}(\mathbf{x} - \mathbf{X}_a) \qquad (5.4)$$

A molecular orbital is then represented by a superposition of the form

$$\varphi_\zeta(\mathbf{x}) = \sum_{nlma} \chi_{nlm}(\mathbf{x} - \mathbf{X}_a)C_{\tau,\zeta} \equiv \sum_\tau \chi_\tau(\mathbf{x})C_{\tau,\zeta} \qquad (5.5)$$

The normalization condition for the molecular orbitals is

$$1 = \int d^3x_j \; \varphi_\zeta^*(\mathbf{x}_j)\varphi_\zeta(\mathbf{x}_j) = \sum_{\tau'}\sum_\tau C_{\tau',\zeta}^* m_{\tau',\tau} C_{\tau,\zeta} \qquad (5.6)$$

where

$$m_{\tau',\tau} \equiv \int d^3x_j \; \chi_{\tau'}^*(\mathbf{x}_j)\chi_\tau(\mathbf{x}_j) \qquad (5.7)$$

is the matrix of many-center Sturmian overlap integrals. The matrix $m_{\tau',\tau}$ may be evaluated using the properties of hyperspherical harmonics, and we will discuss below the details of how this may be done.

Coulomb Sturmian basis functions are discussed in detail in Appendix B. They have exactly the same form as the familiar hydrogenlike atomic orbitals,

$$\chi_{nlm}(\mathbf{x}) = R_{n,l}(r)Y_{lm}(\theta, \phi) \qquad (5.8)$$

except in the radial part, $R_{n,l}(r)$, the factor Z/n is replaced by a constant, k. The first few Coulomb Sturmian radial functions are

$$R_{1,0}(r) = 2k^{3/2}e^{-kr}$$

$$R_{2,0}(r) = 2k^{3/2}(1 - kr)e^{-kr}$$

$$R_{2,1}(r) = \frac{1}{\sqrt{3}} \, 2k^{3/2}kr \, e^{-kr} \tag{5.9}$$

$$R_{3,1}(r) = 2k^{3/3} \left(1 - 2kr + \frac{2k^2r^2}{3} \right) e^{-kr}$$

The reader can verify that these are precisely the same as hydrogenlike atomic orbitals with the replacement $Z/n \to k$. We now substitute the superposition (5.5) into the one-electron Schrödinger equation (5.1). This gives us:

$$\sum_{nlma} \left[-\frac{1}{2}\nabla^2 + \frac{1}{2}k^2 + v(\mathbf{x}) \right] \chi_{nlm}(\mathbf{x} - \mathbf{X}_a)C_{\tau,\zeta}$$

$$\equiv \sum_{\tau} \left[-\frac{1}{2}\nabla^2 + \frac{1}{2}k^2 + v(\mathbf{x}) \right] \chi_\tau(\mathbf{x})C_{\tau,\zeta} = 0 \tag{5.10}$$

with

$$\epsilon_\zeta \equiv -\frac{1}{2}k^2 \tag{5.11}$$

where each of the Coulomb Sturmian atomic orbitals $\chi_\tau(\mathbf{x}) \equiv \chi_{nlm}(\mathbf{x}-\mathbf{X}_a)$ obeys a one-electron Schrödinger equation of the form

$$\left[-\frac{1}{2}\nabla^2 + \frac{1}{2}k^2 - \frac{nk}{|\mathbf{x} - \mathbf{X}_a|} \right] \chi_{nlm}(\mathbf{x} - \mathbf{X}_a) = 0 \tag{5.12}$$

Taking the scalar product of (5.10) with a conjugate function in our basis set, we obtain

$$\sum_{\tau} \int d^3x \chi_{\tau'}^*(\mathbf{x}) \left[-\frac{1}{2}\nabla^2 + \frac{k^2}{2} + v(\mathbf{x}) \right] \chi_\tau(\mathbf{x})C_{\tau,\zeta} = 0 \tag{5.13}$$

With the notation

$$\mathfrak{W}_{\tau',\tau} \equiv -\frac{1}{k} \int d^3x \chi_{\tau'}^*(\mathbf{x})v(\mathbf{x})\chi_\tau(\mathbf{x}) \tag{5.14}$$

and

$$\mathfrak{S}_{\tau',\tau} \equiv \frac{1}{k^2} \int d^3x \chi_{\tau'}^*(\mathbf{x}) \left(-\frac{1}{2}\nabla^2 + \frac{k^2}{2} \right) \chi_\tau(\mathbf{x})$$

$$= \frac{n}{k} \int d^3x \chi_{\tau'}^*(\mathbf{x}) \frac{1}{|\mathbf{x} - \mathbf{X}_a|} \chi_\tau(\mathbf{x}) \tag{5.15}$$

we obtain a secular equation of the form

$$\sum_{\tau} \left[\mathfrak{W}_{\tau',\tau} - k\mathfrak{S}_{\tau',\tau} \right] C_{\tau,\zeta} = 0 \qquad (5.16)$$

We can call $\mathfrak{W}_{\tau',\tau}$ the *Wulfman integrals* to honor the pioneering contributions of Prof. C.E. Wulfman. The integrals $\mathfrak{S}_{\tau',\tau}$ are called *Shibuya-Wulfman integrals* [Avery, 1989], [Avery, 2000], [Avery and Avery, 2006], [Judd, 1975], [Shibuya and Wulfman, 1965], and methods for their evaluation are discussed below. It can be shown [Koga and Matsuhashi, 1987] that the matrix elements of the many-center potential $\mathfrak{W}_{\tau',\tau}$ can be expressed in terms of the Shibuya-Wulfman integrals by means of the sum rule

$$\mathfrak{W}_{\tau',\tau} = \sqrt{\frac{n'n}{Z_{a'} Z_a}} \sum_{\tau''} K_{\tau',\tau''} K_{\tau'',\tau} \qquad (5.17)$$

where

$$K_{\tau',\tau} \equiv \sqrt{\frac{Z_{a'} Z_a}{n'n}} \mathfrak{S}_{\tau',\tau} \qquad (5.18)$$

With the help of this sum rule, we rewrite the secular equations (5.16) as

$$\sum_{\tau} \left[\sum_{\tau''} K_{\tau',\tau''} K_{\tau'',\tau} - kK_{\tau',\tau} \right] C'_{\tau,\zeta} = 0 \qquad (5.19)$$

with

$$C'_{\tau,\zeta} = \sqrt{\frac{Z_a}{n}} C_{\tau,\zeta} \qquad (5.20)$$

Now suppose that we have solved the secular equation

$$\sum_{\tau} \left[K_{\tau',\tau} - k\delta_{\tau',\tau} \right] C'_{\tau,\zeta} \qquad (5.21)$$

The values of k and $C'_{\tau,\zeta}$ thus obtained will also be solutions to (5.19). To see this, we perform the sum over τ'' in (5.19), making use of (5.21):

$$\sum_{\tau} \left[\sum_{\tau''} K_{\tau',\tau''} K_{\tau'',\tau} - kK_{\tau',\tau} \right] C'_{\tau,\zeta}$$

$$= k \sum_{\tau} \left[\sum_{\tau''} K_{\tau',\tau''} \delta_{\tau'',\tau} C'_{\tau,\zeta} - k\delta_{\tau',\tau} \right] C'_{\tau,\zeta} \qquad (5.22)$$

$$= k \sum_{\tau} \left[K_{\tau',\tau} - k\delta_{\tau',\tau} \right] C'_{\tau,\zeta} = 0$$

Since we have two forms of the molecular Sturmian secular equation, (5.16) and (5.21), one might ask which form is the best. The answer is that if the number of basis functions used is small, (5.16) give the most accurate results. However, particularly for small values of the parameter S, equation (5.21) suffers from problems of overcompleteness when the number of basis functions is increased. By contrast, as Monkhorst and Jeziorski have pointed out [Monkhorst and Jeziorski, 1979], equation (5.16) has no such problems, and therefore it is the method of choice when the basis set used is very large or when S is small. We can call the matrix $K_{\tau',\tau}$ the *Koga matrix* to honor the contributions of Prof. T. Koga and his group.

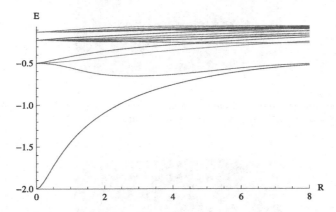

Fig. 5.1 *Energies of the ground state and excited states of H_2^+, calculated by solving equation (5.21). The energies are expressed in Hartrees and are given as a function of the internuclear separation R, expressed in Bohrs.*

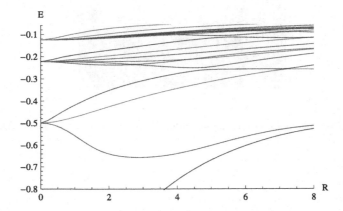

Fig. 5.2 *A closer view of the excited state energies of H_2^+. In the united-atom limit, these energies approach those of the excited states of He^+, i.e., $4/(2n^2)$.*

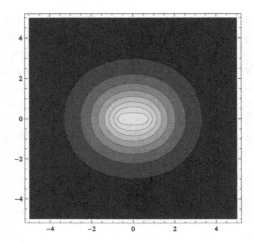

Fig. 5.3 *The ground state molecular orbital of the H_2^+ ion at nuclear separation R=1.21702 Bohrs (S=2, k=1.64335). In the united-atom limit, k=2.*

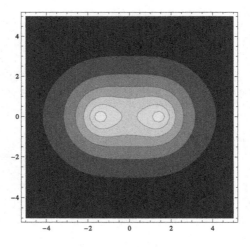

Fig. 5.4 *The same state at nuclear separation 2.98216 Bohrs (S=4, k=1.34131).*

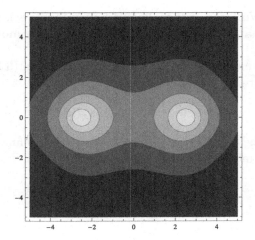

Fig. 5.5 *Here the internuclear distance has been increased to 5.13325 Bohrs (S=6, k=1.16885).*

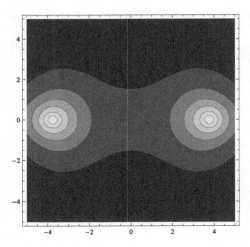

Fig. 5.6 *The same state with nuclear separation 7.50577 Bohrs (S=8, k=1.06585). As the nuclear separation increases, k approaches 1.*

5.2　Shibuya-Wulfman integrals and Sturmian overlap integrals evaluated in terms of hyperpherical harmonics

The Shibuya-Wulfman integrals $\mathfrak{S}_{\tau',\tau}$ defined by equation (5.15) as well as the molecular Sturmian overlap integrals

$$m_{\tau',\tau} \equiv \int d^3x \; \chi_{\tau'}^*(\mathbf{x})\chi_\tau(\mathbf{x}) \tag{5.23}$$

can conveniently be evaluated in reciprocal space. Let us first consider the Sturmian overlap integrals. Using the fact that

$$m_{\tau',\tau} \equiv \int d^3x \; \chi_{\tau'}^*(\mathbf{x})\chi_\tau(\mathbf{x}) = \int d^3p \; \chi_{\tau'}^{t*}(\mathbf{p})\chi_\tau^t(\mathbf{p}) \tag{5.24}$$

where, if we let

$$\mu \equiv (n, l, m) \tag{5.25}$$

while $\tau \equiv (n, l, m, a)$, then

$$
\begin{aligned}
\chi_\tau^t(\mathbf{p}) &\equiv \frac{1}{(2\pi)^{3/2}} \int d^3x \; e^{-i\mathbf{p}\cdot\mathbf{x}}\chi_\tau(\mathbf{x}) = e^{-i\mathbf{p}\cdot\mathbf{X}_a}\chi_\mu^t(\mathbf{p}) \\
\chi_{\tau'}^{t*}(\mathbf{p}) &\equiv \frac{1}{(2\pi)^{3/2}} \int d^3x \; e^{i\mathbf{p}\cdot\mathbf{x}}\chi_{\tau'}^*(\mathbf{x}) = e^{i\mathbf{p}\cdot\mathbf{X}_{a'}}\chi_\mu^{t*}(\mathbf{p})
\end{aligned}
\tag{5.26}
$$

We thus obtain

$$m_{\tau',\tau} = \int d^3p \; e^{i\mathbf{p}\cdot\mathbf{R}}\chi_{\mu'}^{t*}(\mathbf{p})\chi_\mu^t(\mathbf{p}) \tag{5.27}$$

where

$$\mathbf{R} \equiv \mathbf{X}_{a'} - \mathbf{X}_a \tag{5.28}$$

We now make use of V. Fock's relationship [Fock, 1935], [Fock, 1958]

$$\chi_\mu^t(\mathbf{p}) = M(p)Y_{n-1.l.m}(\hat{\mathbf{u}}) \tag{5.29}$$

where

$$M(p) \equiv \frac{4k^{5/2}}{(k^2 + p^2)^2} \tag{5.30}$$

In equation (5.29), $\hat{\mathbf{u}}$ is a 4-dimensional unit vector that defines Fock's projection of momentum space onto the surface of a 4-dimensional hypersphere.

$$\hat{\mathbf{u}} = (u_1, u_2, u_3, u_4) = \left(\frac{2kp_1}{k^2 + p^2}, \frac{2kp_2}{k^2 + p^2}, \frac{2kp_3}{k^2 + p^2}, \frac{k^2 - p^2}{k^2 + p^2} \right) \tag{5.31}$$

while $Y_\mu(\hat{\mathbf{u}})$ is a 4-dimensional hyperspherical harmonic defined by:

$$Y_{\lambda,l,m}(\hat{\mathbf{u}}) = \mathcal{N}_{\lambda,l} C^{1+l}_{\lambda-l}(u_4) Y_{l,m}(u_1, u_2, u_3) \tag{5.32}$$

where

$$\mathcal{N}_{\lambda,l} = (-1)^\lambda i^l (2l)!! \sqrt{\frac{2(\lambda+1)(\lambda-l)!}{\pi(\lambda+l+1)!}} \tag{5.33}$$

is a normalizing constant while

$$C^\alpha_j(u_4) = \sum_{t=0}^{[j/2]} \frac{(-1)^t \Gamma(n+\alpha-t)}{t!(j-2t)!\Gamma(\alpha)} (2u_4)^{j-2t} \tag{5.34}$$

is a Gegenbauer polynomial, and where $Y_{l,m}$ is a familiar 3-dimensional spherical harmonic. The first few hyperspherical harmonics are shown in Table 5.1. The index λ corresponds to $n-1$ so that the Fourier transform of $\chi_{1,0,0}(\mathbf{x})$ is given by $\chi^t_{1,0,0}(\mathbf{p}) = M(p)Y_{0,0,0}(\hat{\mathbf{u}}) = M(p)/(\sqrt{2}\pi)$, and so on. Substituting (5.29) into (5.27), we obtain

$$m_{\tau',\tau} = \int d^3p \, e^{i\mathbf{p}\cdot\mathbf{R}} M(p)^2 Y^*_{n'-1,l',m'}(\hat{\mathbf{u}}) Y_{n-1,l,m}(\hat{\mathbf{u}})$$
$$\equiv \int d^3p \, e^{i\mathbf{p}\cdot\mathbf{R}} M(p)^2 Y^*_{\mu'}(\hat{\mathbf{u}}) Y_\mu(\hat{\mathbf{u}}) \tag{5.35}$$

(Here, and throughout the book, a unit vector is indicated by a "hat".) One can show [Avery, 1989] that the Shibuya-Wulfman integrals can be written in a similar form:

$$\mathfrak{S}_{\tau',\tau} = \int d^3p \, e^{i\mathbf{p}\cdot\mathbf{R}} \left(\frac{2k}{k^2+p^2}\right)^3 Y^*_{\mu'}(\hat{\mathbf{u}}) Y_\mu(\hat{\mathbf{u}}) \tag{5.36}$$

One can also show [Caligiana, 2003] that

$$\int d^3p \, e^{i\mathbf{p}\cdot\mathbf{R}} \left(\frac{2k}{k^2+p^2}\right)^3 Y^*_\mu(\hat{\mathbf{u}}) = (2\pi)^{3/2} f_{n,l}(S) Y_{l,m}(\hat{\mathbf{S}}) \tag{5.37}$$

where $Y_{l,m}$ is an ordinary 3-dimensional spherical harmonic and where

$$\mathbf{S} = \{S_x, S_y, S_z\} \equiv k\mathbf{R} \qquad S = k|\mathbf{R}| \tag{5.38}$$

The function $f_{n,l}(S)$ is defined by

$$k^{3/2} f_{n,l} \equiv R_{n,l} - \frac{1}{2}\sqrt{\frac{(n-l)(n+l+1)}{n(n+1)}} R_{n+1,l}$$
$$- \frac{1}{2}\sqrt{\frac{(n+l)(n-l-1)}{n(n-1)}} R_{n-1,l} \tag{5.39}$$

Table 5.1: The first few 4-dimensional hyperspherical harmonics

λ	l	m	$\sqrt{2}\pi\, Y_{\lambda,l,m}(\mathbf{u})$
0	0	0	1
1	1	1	$i\sqrt{2}(u_1 + iu_2)$
1	1	0	$-i2u_3$
1	1	-1	$-i\sqrt{2}(u_1 - iu_2)$
1	0	0	$-2u_4$

λ	l	m	$\sqrt{2}\pi\, Y_{\lambda,l,m}(\mathbf{u})$
2	2	2	$-\sqrt{3}(u_1 + iu_2)^2$
2	2	1	$2\sqrt{3}u_3(u_1 + iu_2)$
2	2	0	$-\sqrt{2}(2u_3^2 - u_1^2 - u_2^2)$
2	2	-1	$-2\sqrt{3}u_3(u_1 - iu_2)$
2	2	-2	$-\sqrt{3}(u_1 - iu_2)^2$
2	1	1	$-i2\sqrt{3}\, u_4(u_1 + iu_2)$
2	1	0	$2i\sqrt{6}\, u_4 u_3$
2	1	-1	$2i\sqrt{3}\, u_4(u_1 - iu_2)$
2	0	0	$4u_4^2 - 1$

where $R_{n,l}$ is the radial function of the Coulomb Sturmians given in equation (5.9), and where

$$R_{n-1,l} \equiv 0 \qquad \text{if } l > n-1 \qquad (5.40)$$

Similarly, one can show [Caligiana, 2003] that

$$\int d^3p\; e^{i\mathbf{p}\cdot\mathbf{R}} M(p)^2 Y_\mu(\hat{\mathbf{u}}) = (2\pi)^{3/2} g_{n,l}(S) Y_{l,m}(\hat{\mathbf{S}}) \qquad (5.41)$$

where

$$g_{n,l} \equiv f_{n,l} - \frac{1}{2}\sqrt{\frac{(n-l)(n+l+1)}{n(n+1)}} f_{n+1,l}$$
$$- \frac{1}{2}\sqrt{\frac{(n+l)(n-l-1)}{n(n-1)}} f_{n-1,l} \tag{5.42}$$

where we define

$$f_{n-1,l} \equiv 0 \qquad \text{if } l > n-1 \tag{5.43}$$

The first few values of $f_{n,l}(S)$ and $g_{n,l}(S)$ are shown in Table 5.2.

Equations (5.37) and (5.41) are respectively identical with the Shibuya Wulfman integrals and the molecular Sturmian overlap integrals except that they contain only one 4-dimensional hyperspherical harmonic instead of a product of two. Thus the problem of evaluating both $\mathfrak{S}_{\tau',\tau}$ and $m_{\tau',\tau}$ reduces to the problem of evaluating the coefficients

$$c_{\mu'';\mu',\mu} = \int d\Omega_4 Y_{\mu''}^*(\hat{\mathbf{u}}) Y_{\mu'}^*(\hat{\mathbf{u}}) Y_{\mu}(\hat{\mathbf{u}}) \tag{5.44}$$

These coefficients can readily be pre-evaluated once and for all using the hyperangular integration theorems discussed in Appendix C, and they can be stored as a large but very sparse matrix. We then obtain the relationships:

$$Y_{\mu'}^*(\hat{\mathbf{u}}) Y_{\mu}(\hat{\mathbf{u}}) = \sum_{\mu''} Y_{\mu''}(\hat{\mathbf{u}}) c_{\mu'';\mu',\mu} \tag{5.45}$$

$$\mathfrak{S}_{\tau',\tau} = (2\pi)^{3/2} \sum_{\mu''} Y_{l'',m''}(\hat{\mathbf{S}}) f_{n'',l''}(S) c_{\mu'';\mu',\mu} \tag{5.46}$$

and

$$m_{\tau',\tau} = (2\pi)^{3/2} \sum_{\mu''} Y_{l'',m''}(\hat{\mathbf{S}}) g_{n'',l''}(S) c_{\mu'';\mu',\mu} \tag{5.47}$$

Similar methods can be used to calculate the Wulfman integrals $\mathfrak{W}_{\tau',\tau}$ [Avery, 2000]. The first few Shibuya-Wulfman integrals are shown in Table 5.3. We can notice that when $S = 0$ the diagonal elements are 1, while the off-diagonal elements vanish. The first few displaced Coulomb Sturmian overlap integrals $m_{\tau',\tau}$ are shown in Table 5.4.

Table 5.2: $g_{n,l}(S)$ and $f_{n,l}(S)$, where $S \equiv k|\mathbf{X}_{a'} - \mathbf{X}_a|$. The functions $g_{n,l}(S)$ and $f_{n,l}(S)$ appear respectively in the two-center overlap integrals and the Shibuya-Wulfman integrals.

n	l	$g_{n,l}(S)$	$f_{n,l}(S)$
1	0	$\dfrac{e^{-S}\left(3 + 3S + S^2\right)}{3}$	$e^{-S}\left(1 + S\right)$
2	0	$-\dfrac{1}{6}e^{-S}\left(3 + 3S + 2S^2 + S^3\right)$	$-\dfrac{2}{3}e^{-S}S^2$
2	1	$\dfrac{e^{-S}S\left(3 + 3S + S^2\right)}{6\sqrt{3}}$	$\dfrac{2e^{-S}S(1 + S)}{3\sqrt{3}}$
3	0	$\dfrac{1}{15}e^{-S}S^4$	$\dfrac{1}{3}e^{-S}S^2(-2 + S)$
3	1	$-\dfrac{e^{-S}S^3(1 + S)}{15\sqrt{2}}$	$\dfrac{e^{-S}S(1 + S - S^2)}{3\sqrt{2}}$
3	2	$\dfrac{e^{-S}S^2(3 + 3S + S^2)}{15\sqrt{10}}$	$\dfrac{e^{-S}S^2(1 + S)}{3\sqrt{10}}$

Table 5.3: This table shows the first few Shibuya-Wulfman integrals $\mathfrak{S}_{\tau',\tau}$, as functions of $\mathbf{S} \equiv k(\mathbf{X}_{a'} - \mathbf{X}_a)$, with $S \equiv |\mathbf{S}|$ and $\mathbf{S} \equiv (S\sin\theta\cos\phi, S\sin\theta\sin\phi, S\cos\theta)$. The integrals were generated by means of equation (5.46).

τ'	$\tau = (1,0,0,a)$	$\tau = (2,0,0,a)$
$(1,0,0,a')$	$e^{-S}(1+S)$	$-\dfrac{2}{3}e^{-S}S^2$
$(2,0,0,a')$	$-\dfrac{2}{3}e^{-S}S^2$	$\dfrac{1}{3}e^{-S}(3+3S-2S^2+S^3)$
$(2,1,-1,a')$	$-\dfrac{\sqrt{2}}{3}e^{-S}S(1+S)\sin\theta\, e^{i\phi}$	$\dfrac{1}{3\sqrt{2}}e^{-S}S(-1-S+S^2)\sin\theta\, e^{i\phi}$
$(2,1,0,a')$	$-\dfrac{2}{3}e^{-S}S(1+S)\cos\theta$	$\dfrac{1}{3}e^{-S}S(-1-S+S^2)\cos\theta$
$(2,1,1,a')$	$\dfrac{\sqrt{2}}{3}e^{-S}S(1+S)\sin\theta\, e^{-i\phi}$	$-\dfrac{1}{3\sqrt{2}}e^{-S}S(-1-S+S^2)\sin\theta\, e^{-i\phi}$

Table 5.4: The first few overlap integrals $m_{\tau',\tau} \equiv \int d^3x \chi_{\tau'}^*(\mathbf{x})\chi_\tau(\mathbf{x})$ between displaced Coulomb Sturmians. The definitions of S, θ and ϕ are the same as in Table 5.3. The integrals were evaluated by means of equation (5.47).

τ'	$\tau = (1,0,0,a)$	$\tau = (2,0,0,a)$
$(1,0,0,a')$	$\dfrac{1}{3}e^{-S}(3+3S+S^2)$	$-\dfrac{1}{6}e^{-S}(3+3S+2S^2+S^3)$
$(2,0,0,a')$	$-\dfrac{1}{6}e^{-S}(3+3S+2S^2+S^3)$	$\dfrac{1}{15}e^{-S}(15+15S+5S^2+S^4)$
$(2,1,-1,a')$	$-\dfrac{1}{6\sqrt{2}}e^{-S}S(3+3S+S^2)\sin\theta\,e^{i\phi}$	$\dfrac{1}{15\sqrt{2}}e^{-S}S^3(1+S)\sin\theta\,e^{i\phi}$
$(2,1,0,a')$	$-\dfrac{1}{6}e^{-S}S(3+3S+S^2)\cos\theta$	$\dfrac{1}{15}e^{-S}S^3(1+S)\cos\theta$
$(2,1,1,a')$	$\dfrac{1}{6\sqrt{2}}e^{-S}S(3+3S+S^2)\sin\theta\,e^{-i\phi}$	$-\dfrac{1}{15\sqrt{2}}e^{-S}S^3(1+S)\sin\theta\,e^{-i\phi}$

5.3 Molecular calculations using the isoenergetic configurations

We now introduce N-electron configurations which are Slater determinants of the form:

$$|\Phi_\nu\rangle = |\varphi_\zeta \varphi_{\zeta'} \varphi_{\zeta''} \cdots| \equiv \frac{1}{\sqrt{N!}} \begin{vmatrix} \varphi_\zeta(\mathbf{x}_1) & \varphi_{\zeta'}(\mathbf{x}_1) & \varphi_{\zeta''}(\mathbf{x}_1) & \cdots \\ \varphi_\zeta(\mathbf{x}_2) & \varphi_{\zeta'}(\mathbf{x}_2) & \varphi_{\zeta''}(\mathbf{x}_2) & \cdots \\ \varphi_\zeta(\mathbf{x}_3) & \varphi_{\zeta'}(\mathbf{x}_3) & \varphi_{\zeta''}(\mathbf{x}_3) & \cdots \\ \vdots & \vdots & \vdots & \end{vmatrix} \qquad (5.48)$$

where the molecular spin-orbitals $\varphi_\zeta(\mathbf{x})$ satisfy

$$\left[-\frac{1}{2}\nabla_j^2 + \frac{k^2}{2} + \beta_\nu v(\mathbf{x}_j) \right] \varphi_\zeta(\mathbf{x}_j) = 0 \qquad v(\mathbf{x}_j) = \sum_a \frac{Z_a}{|\mathbf{x}_j - \mathbf{X}_a|} \qquad (5.49)$$

Since the individual molecular orbitals satisfy (5.49), the configurations $|\Phi_\nu\rangle$ are solutions to the separable N-electron equation:

$$\sum_{j=1}^N \left[-\frac{1}{2}\nabla_j^2 + \frac{k^2}{2} + \beta_\nu v(\mathbf{x}_j) \right] |\Phi_\nu\rangle = 0 \qquad (5.50)$$

which can also be written in the form:

$$\left[\sum_{j=1}^N \left(-\frac{1}{2}\nabla_j^2 + \frac{k^2}{2} \right) + \beta_\nu V_0(\mathbf{x}) \right] |\Phi_\nu\rangle$$

$$= \left[-\frac{1}{2}\sum_{j=1}^N \nabla_j^2 + \beta_\nu V_0(\mathbf{x}) - E_\kappa \right] |\Phi_\nu\rangle = 0 \qquad (5.51)$$

where $\mathbf{x} \equiv (\mathbf{x}_1, \mathbf{x}_2, \ldots, \mathbf{x}_N)$ and

$$E_\kappa = -\sum_{j=1}^N \frac{k^2}{2} = -\frac{Nk^2}{2} \qquad (5.52)$$

and where

$$V_0(\mathbf{x}) \equiv \sum_{j=1}^N v(\mathbf{x}_j) = \sum_{j=1}^N \sum_a \frac{Z_a}{|\mathbf{x}_j - \mathbf{X}_a|} \qquad (5.53)$$

Comparing (5.51) with (4.6), we can see that they are the same. Thus the isoenergetic solutions to the approximate N-electron Schrödinger equation

(5.51) form a generalized Sturmian basis set. We would like to use these configurations to build up solutions to the N-electron Schrödinger equation

$$\left[\sum_{j=1}^{N}\left(-\frac{1}{2}\nabla_j^2 + \frac{k^2}{2}\right) + V(\mathbf{x})\right]|\Psi_\kappa\rangle = 0 \qquad (5.54)$$

with

$$V(\mathbf{x}) = V_0(\mathbf{x}) + \sum_{i>j}^{N}\frac{1}{r_{ij}} \qquad (5.55)$$

Thus we write

$$|\Psi_\kappa\rangle \approx \sum_{\nu}|\Phi_\nu\rangle B_{\nu\kappa} \qquad (5.56)$$

Substituting this into the N-electron Schrödinger equation, and taking the scalar product with a conjugate configuration, we obtain the secular equations:

$$\sum_{\nu}\langle\Phi_{\nu'}|\left[\sum_{j=1}^{N}\left(-\frac{1}{2}\nabla_j^2 + \frac{k^2}{2}\right) + V(\mathbf{x})\right]|\Phi_\nu\rangle B_{\nu\kappa} = 0 \qquad (5.57)$$

We now introduce a k-independent matrix representing the total potential based on the configurations $|\Phi_\nu\rangle$:

$$T_{\nu'\nu}^{(N)} \equiv -\frac{1}{k}\langle\Phi_{\nu'}|V(\mathbf{x})|\Phi_\nu\rangle \qquad (5.58)$$

and another k-independent matrix

$$\mathfrak{S}_{\nu'\nu}^{(N)} \equiv \frac{1}{k^2}\langle\Phi_{\nu'}^*|\sum_{j=1}^{N}\left(-\frac{1}{2}\nabla_j^2 + \frac{k^2}{2}\right)|\Phi_\nu\rangle \qquad (5.59)$$

In terms of these matrices, the secular equations become:

$$\sum_{\nu}\left[T_{\nu'\nu}^{(N)} - k\mathfrak{S}_{\nu'\nu}^{(N)}\right] B_{\nu\kappa} = 0 \qquad (5.60)$$

Solving equation (5.60), we obtain k for each state κ and thus the energy $E_\kappa = -\frac{Nk^2}{2}$. For a given state κ, the value of k then determines the weighting factors $\beta_{\nu_1}, \beta_{\nu_2}, \ldots$ needed to make each configuration $|\Phi_{\nu_1}\rangle, |\Phi_{\nu_2}\rangle, \ldots$ correspond to the same energy E_κ.

In order to build the N-electron matrices $T_{\nu'\nu}^{(N)}$ and $\mathfrak{S}_{\nu'\nu}^{(N)}$ and solve equation (5.60), we must first obtain the coefficients $C_{\tau\zeta}$ by solving (5.16) or (5.21). In the case of diatomic molecules, we begin by picking a value of the parameter $S = kR$, where R is the interatomic distance and k is the exponent of the Coulomb Sturmian basis set. Neither R nor k is known at this point, but only their product S. As we shall see below, for the diatomic case, all of the integrals involved in equations (5.16) and (5.21) are pure functions of S. Having chosen S, we can thus solve the one-electron secular equations and obtain the coefficients $C_{\tau\zeta}$ and the spectrum of ratios k/β_ν. We are then able to solve equation 5.60, which gives us a spectrum of k-values, and thus energies $-Nk^2/2$, and the eigenvectors $B_{\nu\kappa}$. From a k-value, we also get the unscaled distance $R = S/k$. We repeat the procedure for a range of S-values and interpolate to find the solutions as functions of R.

In the case of polyatomic molecules, one can choose a set of angles between the nuclei; these are left fixed under scaling of the coordinate system. The procedure is then similar to that described for the diatomic case.

5.4 Building $T_{\nu'\nu}^{(N)}$ and $\mathfrak{S}_{\nu'\nu}^{(N)}$ from 1-electron components

We have already discussed how the matrix of many-center Sturmian overlap integrals

$$m_{\tau'\tau} \equiv \int d^3x_j \; \chi_{\tau'}^*(\mathbf{x}_j)\chi_\tau(\mathbf{x}_j) \tag{5.61}$$

may be evaluated using the properties of hyperspherical harmonics (5.47). The matrix $m_{\tau'\tau}$ is needed in order to normalize the molecular orbitals $\varphi_\zeta(\mathbf{x}_j)$, the normalization condition on the coefficients $C_{\tau,\zeta}$ being

$$1 = \int d^3x_j \; \varphi_\zeta^*(\mathbf{x}_j)\varphi_\zeta(\mathbf{x}_j) = \sum_{\tau'}\sum_\tau C_{\tau',\zeta}^* m_{\tau',\tau} C_{\tau,\zeta} \tag{5.62}$$

Having performed the normalization, we then need to transform the nuclear attraction matrix elements $\mathfrak{W}_{\tau'\tau}$ to a representation based on the molecular orbitals:

$$\tilde{v}_{\zeta'\zeta} \equiv \int d^3x_j \; \varphi_{\zeta'}^*(\mathbf{x}_j)v(\mathbf{x}_j)\varphi_\zeta(\mathbf{x}_j) = -k\sum_{\tau'}\sum_\tau C_{\tau'\zeta'}^* \mathfrak{W}_{\tau'\tau} C_{\tau\zeta} \tag{5.63}$$

Once we are in possession of the 1-electron matrix elements $\tilde{v}_{\zeta'\zeta}$, we can evaluate

$$T_{\nu'\nu}^{0,(N)} \equiv -\frac{1}{k}\langle \Phi_{\nu'}^*|V_0(\mathbf{x})|\Phi_\nu\rangle \tag{5.64}$$

by means of the Slater-Condon rules. Because of the potential-weighted orthonormality relations obeyed by generalized Sturmian basis sets (Appendix B), we expect the matrix $T_{\nu'\nu}^{0,(N)}$ to be diagonal. We next transform the matrix of Shibuya-Wulfman integrals to a representation based on the molecular orbitals:

$$\tilde{\mathfrak{S}}_{\zeta'\zeta} \equiv \sum_{\tau'} \sum_{\tau} C_{\tau'\zeta'}^* \mathfrak{S}_{\tau'\tau} C_{\tau\zeta} \tag{5.65}$$

From these 1-electron matrix elements, the N-electron matrix $\mathfrak{S}_{\nu'\nu}^{(N)}$ can be constructed with the help of the Slater-Condon rules. Finally we must deal with the difficult term

$$T_{\nu'\nu}^{\prime(N)} \equiv -\frac{1}{k} \left\langle \Phi_{\nu'}^* \left| \sum_{i>j}^{N} \frac{1}{r_{ij}} \right| \Phi_{\nu} \right\rangle \tag{5.66}$$

which is a k-independent matrix representing the effects of interelectron repulsion, and this will be discussed in the next section.

5.5 Interelectron repulsion integrals for molecular Sturmians from hyperspherical harmonics

We will now show that just as the theory of hyperspherical harmonics can be used to facilitate the calculation of Shibuya-Wulfman and molecular Sturmian overlap integrals, it also provides a method for very rapid calculation of the most important interelectron repulsion integrals involving molecular Sturmians. We again make use of momentum space:

Let $\rho_{\mu_1,\mu_2}(\mathbf{x} - \mathbf{X}_a)$ and $\rho_{\mu_3,\mu_4}(\mathbf{x}' - \mathbf{X}_{a'})$ be two electron density distributions, centered respectively on nuclei at the positions $\mathbf{X}_a$ and $\mathbf{X}_{a'}$. Then the interelectron repulsion between them is given by the integral:

$$J_{\mu_1,\mu_2,\mu_3,\mu_4} = \int d^3x \int d^3x' \rho_{\mu_1,\mu_2}(\mathbf{x} - \mathbf{X}_a) \frac{1}{|\mathbf{x} - \mathbf{x}'|} \rho_{\mu_3,\mu_4}(\mathbf{x}' - \mathbf{X}_{a'}) \tag{5.67}$$

If we introduce the Fourier transform representation

$$\frac{1}{|\mathbf{x} - \mathbf{x}'|} = \frac{1}{2\pi^2} \int d^3p \, \frac{1}{p^2} e^{-i\mathbf{p}\cdot(\mathbf{x}-\mathbf{x}')} \tag{5.68}$$

we can rewrite $J_{\mu_1,\mu_2,\mu_3,\mu_4}$ in the form

$$J_{\mu_1,\mu_2,\mu_3,\mu_4} = 4\pi \int d^3p \, \frac{1}{p^2} e^{i\mathbf{p}\cdot\mathbf{R}} \rho_{\mu_1,\mu_2}^t(\mathbf{p}) \rho_{\mu_3,\mu_4}^t(-\mathbf{p}) \tag{5.69}$$

where $\mathbf{R} = \mathbf{X}_{a'} - \mathbf{X}_a$ and

$$\rho_{\mu_i,\mu_j}^t(\mathbf{p}) = \frac{1}{(2\pi)^{3/2}} \int d^3x \, \rho_{\mu_i,\mu_j}(\mathbf{x}) e^{-i\mathbf{p}\cdot\mathbf{x}} \tag{5.70}$$

Now let $R_{n,l}(2r)$ be a Coulomb Sturmian radial function with r replaced by $2r$, and we let $g(r)$ be any function of r. From the completeness property of Sturmian basis sets we know that it is possible to make an expansion of the form

$$g(r) = \sum_n a_n R_{n,l}(2r) \qquad (5.71)$$

and from the potential-weighted orthonormality relations (B.11) it follows that the expansion coefficients will be given by

$$a_n = \frac{n}{2} \int_0^\infty dr \; r \; R_{n,l}(2r)g(r) \qquad (5.72)$$

We now let the density be composed of a product of two Coulomb Sturmian basis functions:

$$\rho_{\mu_1,\mu_2}(\mathbf{x}) = \chi_{\mu_1}^*(\mathbf{x})\chi_{\mu_2}(\mathbf{x}) = R_{n_1,l_1}(r)R_{n_2,l_2}(r)Y_{l_1,m_1}^*(\hat{\mathbf{x}})Y_{l_2,m_2}(\hat{\mathbf{x}}) \qquad (5.73)$$

If we make the expansion

$$\rho_{\mu_1,\mu_2}(\mathbf{x}) = \sum_{\mu''} R_{n'',l''}(2r)Y_{l'',m''}(\hat{\mathbf{x}}) \, \mathcal{C}_{\mu'';\mu_1,\mu_2} \equiv \sum_{\mu''} \chi_{\mu''}(2\mathbf{x}) \, \mathcal{C}_{\mu'';\mu_1,\mu_2}$$

$$(5.74)$$

then the coefficients in the expansion will be given by

$$\mathcal{C}_{\mu'';\mu_1,\mu_2} = \frac{n''}{2} \int_0^\infty dr \; r \; R_{n'',l''}(2r)R_{n_1,l_1}(r)R_{n_2,l_2}(r)$$

$$\times \int d\Omega_3 \; Y_{l'',m''}^*(\hat{\mathbf{x}})Y_{l_1,m_1}^*(\hat{\mathbf{x}})Y_{l_2,m_2}(\hat{\mathbf{x}}) \qquad (5.75)$$

Like the coefficients $c_{\mu'';\mu',\mu}$, the coefficients $\mathcal{C}_{\mu'';\mu',\mu}$ form a large but very sparse matrix which can be pre-calculated once and for all and stored. The series in (5.74) terminates and the expansion is exact. Making use of the relationships (5.73) and (5.74), we obtain the result

$$J_{\mu_1,\mu_2,\mu_3,\mu_4} = \int d^3x \int d^3x' \rho_{\mu_1,\mu_2}(\mathbf{x} - \mathbf{X}_a)\frac{1}{|\mathbf{x} - \mathbf{x}'|}\rho_{\mu_3,\mu_4}(\mathbf{x}' - \mathbf{X}_{a'})$$

$$= \sum_{\mu',\mu} J_{\mu',\mu}\mathcal{C}_{\mu',\mu_1,\mu_2}\mathcal{C}_{\mu,\mu_3,\mu_4} \qquad (5.76)$$

where

$$J_{\mu',\mu} = 4\pi \int d^3p \frac{1}{p^2}e^{i\mathbf{p}\cdot\mathbf{R}}\rho_{\mu'}^t(\mathbf{p})\rho_\mu^t(-\mathbf{p}) \qquad (5.77)$$

and where

$$\rho_{\mu'}(\mathbf{x}) = R_{n',l'}(2r)Y_{l',m'}(\hat{\mathbf{x}})$$

$$\rho_\mu(\mathbf{x}) = R_{n,l}(2r)Y_{l,m}(\hat{\mathbf{x}}) \qquad (5.78)$$

Then, making use of Fock's relationship we have:

$$
\begin{aligned}
\rho_{\mu'}^{t}(\mathbf{p}) &= \tilde{M}(p)Y_{n'-1,l',m'}(\hat{\mathbf{w}}) \equiv \tilde{M}(p)Y_{\mu'}(\hat{\mathbf{w}}) \\
\rho_{\mu}^{t}(-\mathbf{p}) &= (-1)^{l}\tilde{M}(p)Y_{n-1,l,m}(\hat{\mathbf{w}}) \equiv (-1)^{l}\tilde{M}(p)Y_{\mu}(\hat{\mathbf{w}})
\end{aligned}
\tag{5.79}
$$

where $\tilde{M}(p)$ and $\hat{\mathbf{w}}$ are similar to $M(p)$ and $\hat{\mathbf{u}}$, but with double the k value.

$$
\hat{\mathbf{w}} = (w_1, w_2, w_3, w_4) = \left(\frac{4kp_1}{4k^2 + p^2}, \frac{4kp_2}{4k^2 + p^2}, \frac{4kp_3}{4k^2 + p^2}, \frac{4k^2 - p^2}{4k^2 + p^2} \right) \tag{5.80}
$$

and

$$
\tilde{M}(p) \equiv \frac{4(2k)^{5/2}}{(4k^2 + p^2)^2} \tag{5.81}
$$

Then

$$
J_{\mu',\mu} = 4\pi \int d^3p \frac{1}{p^2} e^{i\mathbf{p}\cdot\mathbf{R}} \tilde{M}^2(p)(-1)^{l} Y_{\mu'}(\hat{\mathbf{w}}) Y_{\mu}(\hat{\mathbf{w}}) \tag{5.82}
$$

Since we know how to represent the product $(-1)^{l}Y_{\mu'}(\hat{\mathbf{w}})Y_{\mu}(\hat{\mathbf{w}})$ in terms of $Y_{\mu''}(\hat{\mathbf{w}})$, we can express the matrix $J_{\mu',\mu}$ in terms of a single vector, $J_{\mu''}$: Let

$$
\tilde{c}_{\mu'';\mu',\mu} \equiv (-1)^{l} \int d\Omega_4 Y_{\mu''}^{*}(\hat{\mathbf{w}}) Y_{\mu'}(\hat{\mathbf{w}}) Y_{\mu}(\hat{\mathbf{w}}) \tag{5.83}
$$

Then

$$
J_{\mu',\mu} = \sum_{\mu''} J_{\mu''} \tilde{c}_{\mu'';\mu',\mu} \tag{5.84}
$$

where

$$
J_{\mu} \equiv 4\pi \int d^3p \frac{1}{p^2} e^{i\mathbf{p}\cdot\mathbf{R}} \tilde{M}^2(p) Y_{\mu}(\hat{\mathbf{w}}) \tag{5.85}
$$

The coefficients $\tilde{c}_{\mu'';\mu',\mu}$ differ slightly from the coefficients $c_{\mu'';\mu',\mu}$, but they too form a large but very sparse matrix that can be pre-calculated and stored. We must now evaluate J_{μ}. To do so, it is convenient to introduce the notation

$$
\rho_{\mu}^{t}(\mathbf{p}) = \tilde{M}(p)Y_{\mu}(\hat{\mathbf{w}}) \equiv i^{-l} R_{n,l}^{t}(p) Y_{l,m}(\hat{\mathbf{p}}) \tag{5.86}
$$

where

$$
R_{n,l}^{t}(p) \equiv \sqrt{\frac{2}{\pi}} \int_{0}^{\infty} dr \; r^2 j_l(pr) R_{n,l}(2r) \tag{5.87}
$$

Then, expanding the plane wave in terms of spherical harmonics and spherical Bessel functions and integrating over the solid angle in momentum space, we have:

$$
\begin{aligned}
J_\mu &= \pi \int d^3p \, \frac{1}{p^2} e^{i\mathbf{p}\cdot\mathbf{R}} \tilde{M}^2(p) Y_\mu(\hat{\mathbf{w}}) \\
&= (4\pi)^2 Y_{l,m}(\hat{\mathbf{S}}) \int_0^\infty dp \, \tilde{M}(p) R_{n,l}^t(p) j_l(pR)
\end{aligned}
\tag{5.88}
$$

The radial p-integrals in equation (5.88) are simple enough to be evaluated exactly by Mathematica, and they depend only on n and l. They can conveniently be stored as interpolation functions. It is also convenient to initialize by performing the sum shown in equation (5.84). This sum, and the sums required for the evaluation of $J_{\mu_1,\mu_2,\mu_3,\mu_4}$ from $J_{\mu',\mu}$ can be performed very rapidly because of the sparseness of the coefficients.

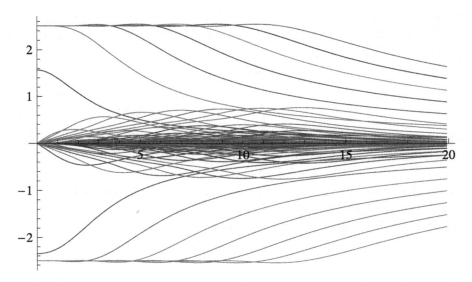

Fig. 5.7   The integrals $\int_0^\infty dp\ \tilde{M}(p)R_{n,l}^t(p)j_l(pR)/k$ of equation (5.88) are shown here plotted as functions of $S \equiv kR$. There are 105 functions, corresponding $n = 1, 2, \ldots, 14$ and $l = 0, 1, \ldots, n-1$.

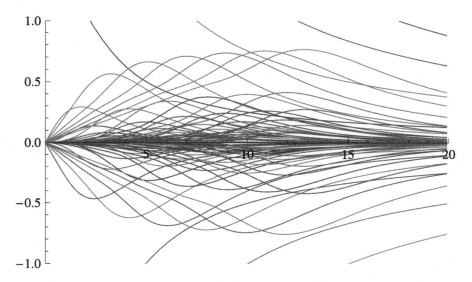

Fig. 5.8   The integrals $\int_0^\infty dp\ \tilde{M}(p)R_{n,l}^t(p)j_l(pR)/k$ shown in more detail. For small values of S the integrals are proportional to S^l, while for large values they are proportional to $1/S^{l+1}$.

5.6 Many-center integrals treated by Gaussian expansions (Appendix E)

One also needs to calculate 3-center and 4-center integrals of the form

$$J_{\tau_1,\tau_2,\tau_3\tau_4} = \int d^3x \int d^3x' \; \chi_{\tau_1}^*(\mathbf{x})\chi_{\tau_2}(\mathbf{x})\frac{1}{|\mathbf{x}-\mathbf{x}'|}\chi_{\tau_3}^*(\mathbf{x}')\chi_{\tau_4}(\mathbf{x}') \quad (5.89)$$

where the centers $\mathbf{X}_{a_1},\ldots,\mathbf{X}_{a_4}$ may in general be 4 different points. Even in this difficult case, molecular Sturmian basis functions have very marked advantages. One can show (Appendix E) that $J_{\tau_1,\tau_2,\tau_3\tau_4}/k$ is independent of k. They may therefore be calculated once and for all and stored. One can also show (Appendix E) that the Coulomb Sturmian atomic orbitals can be expressed in terms of Gaussian expansions, where the Gaussian exponents α_i are universals that need never be changed despite changes in scaling due changes in the value of k. The coefficients $\gamma_{0,i}, \gamma_{1,i}, \ldots$ in the following expression

$$s^j e^{-s} \approx \sum_i \gamma_{j,i} e^{-\alpha_i s^2} \qquad s \equiv kr \qquad (5.90)$$

are also universals, and they too need never be changed, despite changes in scaling. We make the expansion

$$\chi_\tau(\mathbf{x}) = \chi_{n,l,m}(\mathbf{x}-\mathbf{X}_a) \approx k^{3/2} \sum_i \Gamma_{n,l,i} \; e^{-\alpha_i|k\mathbf{x}-k\mathbf{X}_a|^2} R_l^m(k\mathbf{x}-k\mathbf{X}_a)$$

where the coefficients $\Gamma_{n,l,i}$ are defined by the relationship

$$\sqrt{\frac{2l+1}{4\pi}}\tilde{R}_{n,l}(s)s^{-l} \approx \sum_i \Gamma_{n,l,i} e^{-\alpha_i s^2} \qquad (5.91)$$

with $\tilde{R}_{n,l}(s) \equiv R_{n,l}(r)/k^{3/2}$ and where R_l^m is a regular solid harmonic (Appendix E). Figure 5.9 shows the Gaussian expansion of $e^{-s} \equiv e^{-kr}$, while the table shows coefficients in the expansion.

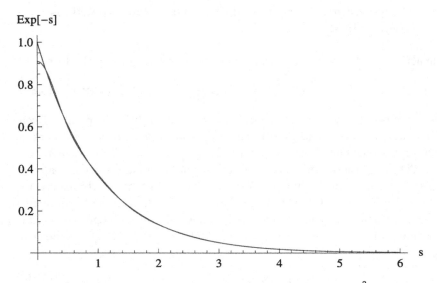

Fig. 5.9   *This figure shows the Gaussian expansion $e^{-s} \approx \sum_i \gamma_{0,i} e^{-\alpha_i s^2}$, using the 10 coefficients and exponents shown in Table 5.5. The expansion is reasonably accurate throughout most of the range, but it fails to produce the sharp cusp near $s = 0$.*

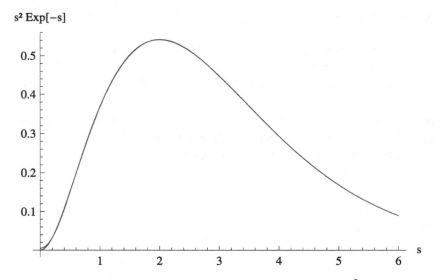

Fig. 5.10   *Here we see the Gaussian expansion $s^2 e^{-s} \approx \sum_i \gamma_{2,i} e^{-\alpha_i s^2}$. As in Figure 5.9, the expansion is compared with the exact function.*

Table 5.5: Universal coefficients for Gaussian expansions of Coulomb Sturmians: They are used in the relationship $s^j e^{-s} = \sum_i \gamma_{j,i} e^{-\alpha_i s^2}$, where $s \equiv kr$. When k changes with scaling, the Gaussian expansion changes scale automatically.

i	α_i	$\gamma_{0,i}$	$\gamma_{1,i}$	$\gamma_{2,i}$
1	5.12	0.474589	−0.456553	−0.011253
2	2.56	−0.409842	0.420846	−0.135640
3	1.28	0.522704	−0.461490	−0.030952
4	0.64	−0.028869	0.157189	−0.390496
5	0.32	0.237377	0.008340	−0.284720
6	0.16	0.074194	0.248277	0.001174
7	0.08	0.039810	0.147977	0.631545
8	0.04	−0.001091	0.025882	0.224411
9	0.02	0.000808	−0.001018	0.000462
10	0.01	−0.000129	0.000170	0.000468

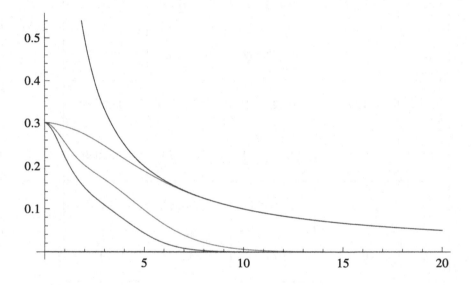

Fig. 5.11 *Interelectron repulsion integrals $J_{\tau_1,\tau_2,\tau_3,\tau_4}/k$ for diatomic molecules for $\mathbf{n} =$
$(2,2,2,2)$ and $\mathbf{l} = (0,0,0,0)$ as functions of $S = kR$. The lowest curve shows a the
results when $(a_1, a_2, a_3, a_4) = (S.0, S, 0)$. The next higher curve shows the case where
$(a_1, a_2, a_3, a_4) = (S.S, S, 0)$. The highest curve, which is the $(a_1, a_2, a_3, a_4) = (S.S, 0, 0)$
case, is compared with $1/S$, which it approaches asymptoticly as S becomes large.*

5.7 A pilot calculation

We have made a small pilot calculation on the dissociation of the hydrogen molecule, using a very restricted basis set. In the calculation shown in Fig-

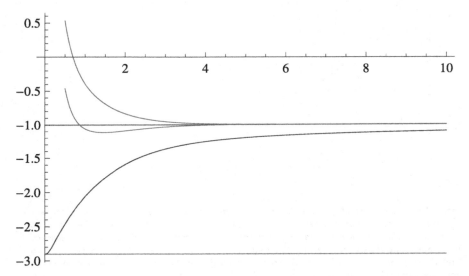

Fig. 5.12 *This figure shows the results of a preliminary calculation on the dissociation of the hydrogen molecule using a very restricted basis set. Energies are shown in Hartrees as functions of the internuclear separation, measured in Bohrs. The lowest curve shows the ground-state electronic energy by itself, without internuclear repulsion. The two upper curves show the ground state and first excited singlet state electronic energies with nuclear repulsion added, i.e. the total energies of the two states. The calculated equilibrium bond length is 1.41 Bohrs, which can be compared with the experimental value, 1.40 Bohrs. It can be seen from the figure that at a separation of 5 Bohrs or more, the molecule is completely dissociated, and in fact the calculated wave function at that internuclear separation corresponds to two neutral hydrogen atoms, each with its own electron, while the total energy corresponds to that of two isolated hydrogen atoms.*

ure 5.12, the ground state wave function changes character as a function of the internuclear separation, R. As $R \to 0$, the wave function becomes more and more dominated by a configuration which is built from two *gerade* molecular orbitals. But as the molecule dissociates, the wave function becomes the linear combination of configurations representing two isolated neutral hydrogen atoms, each with its own electron, and the total energy corresponds to that of two isolated hydrogen atoms.

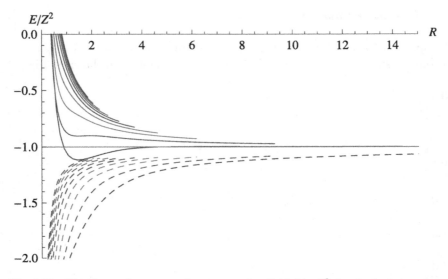

Fig. 5.13   *This figure shows ground-state energies divided by Z^2 for the 2-electron iso-electronic series for homonuclear diatomic molecules, Z being the nuclear charges. The energies in Hartrees are shown as functions of the interatomic distance R, measured in Bohrs. The dotted curves are electronic energies alone, while the solid curves also include internuclear repulsion. For both the solid and dotted curves the lowest curve corresponds to Z=1, the next lowest to Z = 2, and so on. As in Figure 5.12, a very restricted basis set was used for the calculation.*

5.8   Automatic generation of symmetry-adapted basis functions

In Chapter 4, Section 4.2, we discussed a large-Z approximation that could be used when the Generalized Sturmian method is applied to atoms. In the large-Z approximation, the basis set used to treat a particular state is restricted to the set of configurations that become degenerate if interelectron repulsion is completely neglected. We also saw in Chapter 4 that such a set of states could be used as an invariant subset, and thus be used to generate symmetry-adapted basis functions for more accurate calculations where the large-Z approximation is abandoned.

If we compare equation (5.51) with equation (4.6) of Chapter 4, we can see that the two equations are closely analogous. In each case, the configuration $|\Phi_\nu\rangle$ satisfies an approximate Schrödinger equation with a weighted potential V_0, the weighting factors being chosen in such a way as to make all of the configurations in the basis set isoenergetic. Thus both the Goscinskian configurations of Chapter 4 and the isoenergetic configurations of Chapter 5 are examples of generalized Sturmians, as discussed in Appendix B. Therefore it is interesting to ask whether something analogous to the large-Z approximation exists in the case of molecules. What happens if we decide to use as a basis only those molecular configurations that become degenerate if we completely neglect interelectron repulsion? Let us suppose that this degeneracy is not accidental, but is a due degeneracy. It then follows that any set of configurations that become degenerate if interelectron repulsion is completely neglected is closed under the operations of the symmetry group of the molecule, and it can be used as an invariant subset for the automatic generation of symmetry-adapted basis functions needed in a large and accurate calculation. The method can thus be summarized as follows:

(1) Construct the matrices $T_{\nu'\nu}^{(0,N)} \equiv -\frac{1}{k}\langle \Phi_{\nu'}|V_0(\mathbf{x})|\Phi_\nu\rangle$ and $\mathfrak{S}_{\nu'\nu}^{(N)}$, based on configurations that are solutions to (5.51). These will already be diagonal, as was discussed above.

(2) Search for configurations corresponding to the same value of $T_{\nu,\nu}^{(0,N)}/\mathfrak{S}_{\nu,\nu}^{(N)}$. Such a set of degenerate configurations is an invariant subset provided that the search has been sufficiently complete.

(3) For each invariant subset, solve equation (5.60). The eigenfunctions will then be the symmetry-adapted configurations needed for a larger and more accurate calculation.

Chapter 6

AN EXAMPLE FROM ACOUSTICS

6.1 The Helmholtz equation for a non-uniform medium

If $v(\mathbf{x})$ represents the velocity of sound in a non-uniform medium as a function of the position $\mathbf{x}$, a sound-wave of frequency ω_κ satisfies the wave equation

$$\left[\nabla^2 - \frac{1}{v(\mathbf{x})^2}\frac{\partial^2}{\partial t^2}\right]|\Psi_\kappa(\mathbf{x})\rangle e^{\pm i\omega_\kappa t} \tag{6.1}$$

Separating out the time-dependence, we obtain the Helmholtz equation

$$\left[\nabla^2 + \left(\frac{\omega_\kappa}{v(\mathbf{x})}\right)^2\right]|\Psi_\kappa(\mathbf{x})\rangle = 0 \tag{6.2}$$

where $|\Psi_\kappa\rangle$ represents the amplitude of the sound wave as a function of position.

6.2 Homogeneous boundary conditions at the surface of a cube

As an example of how the Helmholtz equation may be solved with the help of symmetry-adapted basis functions, we can consider a cubical cavity, where L is the length of the edges of the cube. The walls of the cube are assumed to be so rigid that it is realistic to require that the amplitude function $|\Psi_\kappa\rangle$ should vanish on the cube's surface. If the Helmholtz equation is difficult to solve because of the non-uniformity of the medium within the cavity, we may nevertheless construct approximate solutions from linear combinations of basis functions:

$$|\Psi_\kappa\rangle = \sum_\nu |\Phi_\nu\rangle C_{\nu,\kappa} \tag{6.3}$$

If we place the origin of our coordinate system at the center of the cubical cavity[1], and impose homogeneous boundary conditions at the surfaces where $x_j = \pm L/2$, the basis functions

$$|\Phi_\nu\rangle = \left(\frac{2}{L}\right)^{3/2} \prod_{j=1}^{3} \sin\left(\frac{\pi n_j x_j}{L} - \frac{n_j \pi}{2}\right) \qquad n_j = 1, 2, 3, \dots \qquad (6.4)$$

obey the same boundary conditions as the desired solution. These basis functions are the harmonics of the cavity for the case where the medium within it is uniform. In equation (6.4), ν denotes the set of three positive integers that characterizes each of the basis functions.

$$\nu = (n_1, n_2, n_3) \qquad (6.5)$$

Since

$$\frac{2}{L} \int_{-L/2}^{L/2} dx_j \, \sin\left(\frac{\pi n_j' x_j}{L} - \frac{n_j' \pi}{2}\right) \sin\left(\frac{\pi n_j x_j}{L} - \frac{n_j \pi}{2}\right) = \delta_{n_j', n_j} \qquad (6.6)$$

it follows that the basis functions are orthonormal,

$$\langle \Phi_{\nu'} | \Phi_\nu \rangle = \delta_{\nu', \nu} \qquad (6.7)$$

Substituting (6.3) into the Helmholtz equation (6.2), we obtain

$$\sum_{\nu=1}^{M} \left[\nabla^2 + \left(\frac{\omega_\kappa}{v(\mathbf{x})}\right)^2\right] |\Phi_\nu\rangle C_{\nu\kappa} = 0 \qquad (6.8)$$

Applying the Laplacian operator to one of our basis functions, we obtain

$$\nabla^2 |\Phi_\nu\rangle = -\left(\frac{\pi}{L}\right)^2 (n_1^2 + n_1^2 + n_3^2)|\Phi_\nu\rangle \qquad (6.9)$$

Making use of (6.9), rearranging the terms in (6.8), and taking the scalar product with a conjugate basis function, we have

$$\sum_{\nu=1}^{M} \langle \Phi_{\nu'} | \left[\left(\frac{\omega_\kappa}{v(\mathbf{x})}\right)^2 - \left(\frac{\pi}{L}\right)^2 (n_1^2 + n_1^2 + n_3^2)\right] |\Phi_\nu\rangle C_{\nu\kappa} = 0 \qquad (6.10)$$

Using the orthonormality of the basis functions, and introducing the definitions

$$\lambda_\kappa \equiv \omega_\kappa^{-2} \qquad T \equiv v(\mathbf{x})^{-2} \qquad (6.11)$$

we can rewrite (6.10) in the form

$$\sum_{\nu=1}^{M} \left[\langle \Phi_{\nu'} | T | \Phi_\nu \rangle - \lambda_\kappa \left(\frac{\pi}{L}\right)^2 (n_1^2 + n_1^2 + n_3^2)\delta_{\nu', \nu}\right] C_{\nu\kappa} = 0 \qquad (6.12)$$

---

[1]Our reason for doing this, instead of placing the origin at a corner of the cube, is to simplify the integrals in the particular example that will be discussed below.

Equation (6.12) still does not have the desired form, (1.3), but we can bring it into this form by renormalizing the basis functions. If we let

$$\mathcal{N}_\nu^2 \equiv \left(\frac{\pi}{L}\right)^2 (n_1^2 + n_1^2 + n_3^2) \qquad |\tilde{\Phi}_\nu\rangle \equiv \frac{1}{\mathcal{N}_\nu}|\Phi_\nu\rangle \qquad (6.13)$$

then (6.12) can be written as

$$\sum_{\nu=1}^{M} \left[\langle\tilde{\Phi}_{\nu'}|T|\tilde{\Phi}_\nu\rangle - \lambda_\kappa \delta_{\nu',\nu}\right] C_{\nu\kappa} = 0 \qquad (6.14)$$

which is identical with (1.3) provided that we let

$$\langle\tilde{\Phi}_{\nu'}|T|\tilde{\Phi}_\nu\rangle \equiv T_{\nu',\nu} \qquad (6.15)$$

6.3 Spherical symmetry of $v(x)$; nonseparability of the Helmholtz equation

To carry the solution further, we now consider the particular example where the velocity of sound throughout the cubic cavity is v_1, except within a spherical volume at the center, of radius R, within which the velocity is v_2. In other words,

$$v(\mathbf{x}) = v(r) = \begin{cases} v_1 & \text{if } r < R \\ \\ v_2 & \text{if } r > R \end{cases} \qquad (6.16)$$

where

$$r \equiv \sqrt{x_1^2 + x_2^2 + x_3^2} \qquad (6.17)$$

Then the Helmholtz equation is non-separable, and our method of solving it by means of a basis set is justified. To evaluate the matrix elements of T, we need to evaluate

$$T_{\nu',\nu} = \langle\tilde{\Phi}_{\nu'}|v(\mathbf{x})^{-2}|\tilde{\Phi}_\nu\rangle = \frac{1}{\mathcal{N}_{\nu'}\mathcal{N}_\nu} \left[v_2^{-2}\delta_{\nu',\nu} + (v_1^{-2} - v_2^{-2})\left(\frac{2}{L}\right)^3 I\right] \qquad (6.18)$$

where

$$I = \int_0^R r^2 dr \int d\Omega \prod_{j=1}^{3} \sin\left(\frac{\pi n_j' x_j}{L} - \frac{n_j'\pi}{2}\right) \sin\left(\frac{\pi n_j x_j}{L} - \frac{n_j\pi}{2}\right) \qquad (6.19)$$

The integral shown in (6.19) can be solved by decomposing the product of basis functions into plane waves: We first note that

$$
\sin\left(\frac{\pi n_j' x_j}{L} - \frac{n_j'\pi}{2}\right) \sin\left(\frac{\pi n_j x_j}{L} - \frac{n_j\pi}{2}\right)
$$

$$
= \frac{1}{4}\left[-e^{-i(n_j'+n_j)\pi/2}e^{i(n_j'+n_j)\pi x_j/L} + e^{-i(n_j'-n_j)\pi/2}e^{i(n_j'-n_j)\pi x_j/L}\right] \quad (6.20)
$$

$$
+ \text{ complex conjugate}
$$

Then, taking the product of three such terms, we have

$$
\prod_{j=1}^{3} \sin\left(\frac{\pi n_j' x_j}{L} - \frac{n_j'\pi}{2}\right) \sin\left(\frac{\pi n_j x_j}{L} - \frac{n_j\pi}{2}\right) = \frac{1}{64}\sum_{i=1}^{64} \varphi_i \, e^{i\mathbf{k}_i \cdot \mathbf{x}} \quad (6.21)
$$

where φ_i is a phase factor. We next recall the expansion of a plane wave in terms of spherical Bessel functions and spherical harmonics:

$$
e^{i\mathbf{k}_i \cdot \mathbf{x}} = 4\pi \sum_{l=0}^{\infty} i^l j_l(k_i r) \sum_{m=-l}^{l} Y_{l,m}^*(\hat{\mathbf{k}}_i) Y_{l,m}(\hat{\mathbf{x}}) = j_0(k_i r) + \cdots \quad (6.22)
$$

When the integral over solid angle is performed, only the first term in the expansion survives:

$$
\int d\Omega \, e^{i\mathbf{k}_i \cdot \mathbf{x}} = 4\pi j_0(k_i r) \quad (6.23)
$$

We thus obtain the expression:

$$
I = \frac{4\pi}{64}\sum_{i=1}^{64} \varphi_i \int_0^R r^2 dr \, j_0(k_i r) = \frac{4\pi}{8}\sum_{i=1}^{8} a_i \int_0^R r^2 dr \, j_0(k_i r) \quad (6.24)
$$

where we have made use of the fact that some of the k_i's are equal to each other and where

$$
k_i = \frac{\pi}{L}\sqrt{\sum_{j=1}^{3}(n_j + s_{i,j}n_j')^2} \qquad i = 1, 2, \ldots, 8 \quad (6.25)
$$

with the signs $s_{i,j}$ defined by

$$
s_{i,j} \equiv \begin{pmatrix} -1 & -1 & -1 \\ -1 & 1 & 1 \\ 1 & -1 & 1 \\ 1 & 1 & -1 \\ 1 & -1 & -1 \\ -1 & 1 & -1 \\ -1 & -1 & 1 \\ 1 & 1 & 1 \end{pmatrix} \quad (6.26)
$$

The constants a_i that appear in equation (6.24) are given by

$$a_i = \begin{cases} \prod_{j=1}^{3} \cos\left[(n_j + s_{i,j}n'_j)\dfrac{\pi}{2}\right] & i = 1,2,3,4 \\[2em] -\prod_{j=1}^{3} \cos\left[(n_j + s_{i,j}n'_j)\dfrac{\pi}{2}\right] & i = 5,6,7,8 \end{cases} \qquad (6.27)$$

The radial integral is an elementary one

$$\int_0^R r^2 dr \, j_0(kr) = \frac{1}{k}\int_0^R r \, dr \, \sin(kr) \qquad (6.28)$$

which yields the result

$$\int_0^R r^2 dr \, j_0(kr) = \begin{cases} \frac{\sin(kR)}{k^3} - \frac{R\cos(kR)}{k^2} & k > 0 \\[1.2em] \frac{R^3}{3} & k = 0 \end{cases} \qquad (6.29)$$

6.4   Diagonalization of invariant blocks

The largest group of symmetry operation that leaves the system (including the boundary conditions) invariant is the cubic or octahedral group $\mathcal{O}_h$:

$$\mathcal{G} = \{\mathcal{R}_1, \mathcal{R}_2, \ldots\} = \mathcal{O}_h \qquad (6.30)$$

Now let us consider the case where the medium inside the cubical cavity is uniform. In that case each of the basis functions will be an exact solution to the eigenvalue problem (6.14). The operations of $\mathcal{O}_h$ can only mix degenerate eigenfunctions, but in the case of a uniform medium, the degenerate eigenfunctions are the basis functions $|\Phi_{n_1,n_2,n_3}\rangle$ corresponding to a particular value of

$$|\nu|^2 \equiv n_1^2 + n_2^2 + n_3^2, \qquad (6.31)$$

i.e. a particular value of the sum of the squares of the three quantum numbers. When $\nu = (1,1,1)$ and $|\nu|^2 = 3$, the solution is non-degenerate. When $|\nu|^2 = 6$ or $|\nu|^2 = 9$, there is a threefold degeneracy, and so on. If the large basis set is chosen by specifying a maximum value of $|\nu|^2$ so that

$$|\nu|^2 = 3, 6, 9, 11, 12, \ldots, |\nu|^2_{max} \qquad (6.32)$$

and

$$\sum \deg(|\nu|) = M \qquad (6.33)$$

then the part of Hilbert space spanned by the large basis set will be invariant under the operations of $\mathcal{O}_h$. Similarly, the subspaces spanned by the subsets

$$W_1 = \text{span} \left\{ |\Phi_{1,1,1}\rangle \right\}$$
$$W_2 = \text{span} \left\{ |\Phi_{2,1,1}\rangle, |\Phi_{1,2,1}\rangle, |\Phi_{1,1,2}\rangle \right\}$$
$$W_3 = \text{span} \left\{ |\Phi_{1,2,2}\rangle, |\Phi_{2,1,2}\rangle, |\Phi_{2,2,1}\rangle \right\} \tag{6.34}$$
$$\vdots$$

will also be invariant under $\mathcal{O}_h$. The sum of the dimensions of the subspaces is of course equal to the dimension of the large basis set:

$$m_1 + m_2 + m_3 + \cdots = 1 + 3 + 3 + \cdots = M \tag{6.35}$$

Figures 6.1 and 6.2 show examples of symmetry-adapted basis functions constructed in this way. In the example shown in these figures, L is chosen to be 1, while $R = 0.2L$. Figure 6.1 shows the symmetry-adapted basis function found by diagonalizing the small block of $T_{\nu',\nu}$ corresponding to $n_1^2 + n_2^2 + n_3^2 = 24$. This basis function penetrates very little into the region occupied by the inner sphere. The opposite is true of the symmetry-adapted basis function shown in Figure 6.2, most of whose amplitude is within the inner sphere. This basis function corresponds to the small invariant block corresponding to $n_1^2 + n_2^2 + n_3^2 = 1$.

Finally, in Figure 6.3, we see a solution to the Helmholtz equation. The basis functions used to obtain this solution were symmetry-adapted basis functions of the type discussed above, and in terms of them, the large $M \times M$ matrix $T_{\nu',\nu}$ is block-diagonal. In the solution to the Helmholtz equation shown in Figure 6.3, the boundary to the inner sphere can clearly be seen.

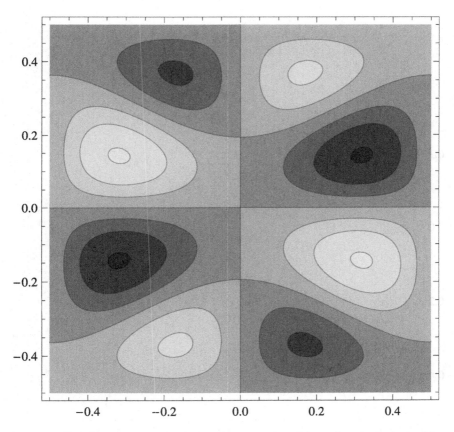

Fig. 6.1   *This figure shows one of the symmetry-adapted basis functions obtained by diagonalizing the matrix $T_{\nu',\nu}$ for the invariant block corresponding to $n_1^2 + n_2^2 + n_3^2 = 24$. The side of the box is chosen to be $L = 1$, and the radius of the central sphere is $R = 0.2$. The basis function is shown in the plane $x_3 = 0.3$, as a function of x_1 and x_2. It penetrates very little into the region of the central sphere. It vanishes on the three planes $x_1 = 0$, $x_2 = 0$ and $x_3 = 0$. When expressed in terms of the symmetry-adapted basis functions, the matrix $T_{\nu',\nu}$ is block diagonal.*

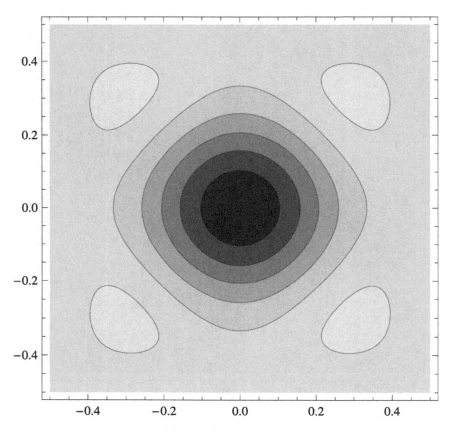

Fig. 6.2   *This figure shows the symmetry-adapted basis function corresponding to the invariant block with $n_1^2 + n_2^2 + n_3^2 = 1$, i.e., it corresponds to the invariant subset W_1. In contrast to the basis function shown in Figure 8.1, it is concentrated almost entirely within the region of the central sphere.*

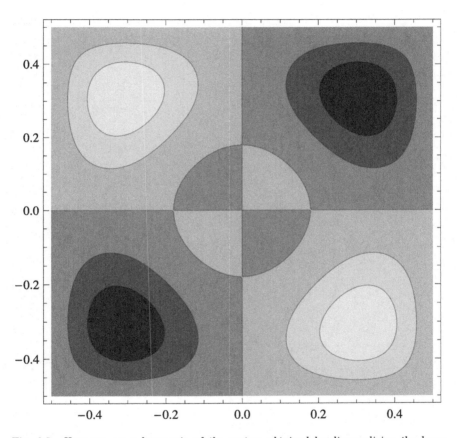

Fig. 6.3 *Here we see a harmonic of the system obtained by diagonalizing the large matrix $T_{\nu',\nu}$. When symmetry-adapted basis functions are used as a basis, this large matrix is block-diagonal. In the harmonic shown here (plotted in the $x_3 = 0$ plane as a function of x_1 and x_2) the surface of the central sphere can be clearly seen.*

Chapter 7

AN EXAMPLE FROM HEAT CONDUCTION

7.1 Inhomogeneous media

The partial differential equation describing heat conduction in an inhomogeneous medium (in the absence of heat sources) has the form:

$$\left[f(\mathbf{x}) \sum_{i,j=1}^{3} \frac{\partial}{\partial x_i} a_{i,j}(\mathbf{x}) \frac{\partial}{\partial x_j} - \frac{\partial}{\partial t} \right] |\Psi_\kappa(\mathbf{x}, t)\rangle = 0 \qquad (7.1)$$

Here $\Psi_\kappa(\mathbf{x}, t)$ represents the temperature at the point $\mathbf{x}$ relative to a standard temperature, $a_{i,j}(\mathbf{x})$ is a position-dependent tensor representing the thermal conductivity of the medium, while $f(\mathbf{x})$ is the inverse specific heat of the medium multiplied by its density. We can try to build up solutions to equation (7.1) from a superposition of the form

$$|\Psi_\kappa(\mathbf{x}, t)\rangle = e^{-\lambda_\kappa t} \sum_{\nu=1}^{M} |\Phi_\nu(\mathbf{x})\rangle C_{\nu,\kappa} \equiv e^{-\lambda_\kappa t} |\Psi_\kappa(\mathbf{x}, 0)\rangle \qquad (7.2)$$

Substituting (7.2) into (7.1) yields

$$\sum_{\nu=1}^{M} \left[f(\mathbf{x}) \sum_{i,j=1}^{3} \frac{\partial}{\partial x_i} a_{i,j}(\mathbf{x}) \frac{\partial}{\partial x_j} + \lambda_\kappa \right] |\Phi_\nu\rangle C_{\nu,\kappa} = 0 \qquad (7.3)$$

If we make the identification

$$T \equiv -f(\mathbf{x}) \sum_{i,j=1}^{3} \frac{\partial}{\partial x_i} a_{i,j}(\mathbf{x}) \frac{\partial}{\partial x_j} \qquad (7.4)$$

and if we assume that our basis functions form an orthonormal set so that

$$\langle \Phi_{\nu'} | \Phi_\nu \rangle = \delta_{\nu',\nu} \qquad (7.5)$$

then (7.3) can be written in the form:

$$\sum_{\nu=1}^{M} \left[T_{\nu',\nu} - \lambda_\kappa \delta_{\nu',\nu} \right] C_{\nu,\kappa} = 0 \qquad (7.6)$$

where

$$T_{\nu',\nu} \equiv \langle \Phi_{\nu'} | T | \Phi_\nu \rangle \qquad (7.7)$$

7.2 A 1-dimensional example

The following very simple example can serve to illustrate the way in which the initial conditions can be matched by a superposition of solutions to equation (7.1): Let us consider a 1-dimensional case where

$$T = -\alpha \frac{d^2}{dx^2} \qquad (7.8)$$

where α is a constant, i.e. where the medium is homogeneous. We can try to build up solutions to (7.6) by a superposition of basis functions of the form:

$$|\Phi_n\rangle = \sqrt{\frac{2}{L}} \sin \left(\frac{\pi n x}{L} \right) \qquad n = 1, 2, 3, \ldots, M \qquad (7.9)$$

Then

$$
\begin{aligned}
T_{n',n} &= -\alpha \langle \Phi_{n'} | \frac{d^2}{dx^2} | \Phi_n \rangle \\
&= \left(\frac{\pi n}{L} \right)^2 \alpha \, \langle \Phi_{n'} | \Phi_n \rangle \qquad (7.10) \\
&= \left(\frac{\pi n}{L} \right)^2 \alpha \, \delta_{n',n}
\end{aligned}
$$

where

$$\langle \Phi_{n'} | \Phi_n \rangle \equiv \frac{2}{L} \int_0^L dx \, \sin \left(\frac{\pi n' x}{L} \right) \sin \left(\frac{\pi n x}{L} \right) = \delta_{n',n} \qquad (7.11)$$

so that equation (7.6) becomes

$$\sum_{n=1}^{M} \left[\left(\frac{\pi n}{L} \right)^2 \alpha \, \delta_{n',n} - \lambda_\kappa \delta_{n',n} \right] C_{n,\kappa} = 0 \qquad (7.12)$$

Since $T_{n',n}$ is already diagonal, the basis functions shown in (7.9) are solutions, and the secular equations simply require that

$$\lambda_\kappa = \left(\frac{\pi n}{L} \right)^2 \alpha \qquad (7.13)$$

The time-dependent solutions then become:

$$|\Psi_n(x,t)\rangle = \sqrt{\frac{2}{L}} \sin\left(\frac{\pi n x}{L}\right) \, \exp\left[-\left(\frac{\pi n}{L}\right)^2 \alpha t\right] \qquad (7.14)$$

Let us suppose that at the time $t = 0$ the temperature distribution is given by

$$|\Psi(x,0)\rangle = g(x) \qquad (7.15)$$

To match this initial condition, we require that

$$g(x) = \sum_n \sqrt{\frac{2}{L}} \sin\left(\frac{\pi n x}{L}\right) a_n \qquad (7.16)$$

The coefficients in the Fourier series are given by

$$a_n = \int_0^L dx \, g(x) \sqrt{\frac{2}{L}} \sin\left(\frac{\pi n x}{L}\right) \qquad (7.17)$$

Then

$$|\Psi(x,t)\rangle = \sum_{n=0}^{M} |\Psi_n(x,t)\rangle a_n \qquad (7.18)$$

The physical situation to which this simple example corresponds might be a metal rod of length L with both ends held at the standard temperature. It is then locally warmed, and as time progresses, the excess temperature distribution evolves in such a way that it becomes progressively more delocalized. An example is shown in Figure 7.1.

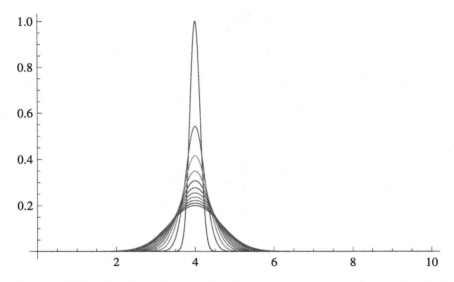

Fig. 7.1 *This figure shows the time-dependence of the excess temperature $\Psi(x,t)$ of equation (7.18), starting with the initial distribution $\Psi(x,0) = \exp[-30(x-4)^2]$. In this illustrative example we have chosen $L = 10$, $M = 100$ and $\alpha = .01$. The times corresponding to the various curves are $t = 0, 2, 4, \ldots, 20$.*

7.3 Heat conduction in a 3-dimensional inhomogeneous medium

In an exactly similar way, solutions to the secular equations (7.6) can be used to build up time-dependent solutions to (7.1) in such a way as to match the initial conditions. Suppose that the initial conditions require that

$$|\Psi(\mathbf{x},0)\rangle = g(\mathbf{x}) \tag{7.19}$$

Then

$$|\Psi(\mathbf{x},t)\rangle = \sum_{\kappa=1}^{M} e^{-\lambda_\kappa t}|\Psi_\kappa(\mathbf{x},0)\rangle a_\kappa \tag{7.20}$$

where $\Psi_\kappa(\mathbf{x})$ is an eigenfunction of the secular equation, (7.6), and where

$$a_\kappa = \langle\Psi_\kappa|g\rangle \tag{7.21}$$

As a simple example, let us think of a cubic volume of side L, inside which there is an inhomogeneous medium, with the walls of the cube held continuously at a standard temperature. Then it is appropriate to use a set of basis functions of the form:

$$|\Phi_\mathbf{n}\rangle = \left(\frac{2}{L}\right)^{3/2} \prod_{j=1}^{3} \sin\left(\frac{\pi n_j x_j}{L}\right) \qquad n_j = 1, 2, 3, \ldots, M^{1/3} \tag{7.22}$$

where we now have placed the origin of our coordinates at one corner of the cube. (In Chapter 6 we placed the origin at the cube's center.) For simplicity, let us suppose that the matrix $a_{i,j}(\mathbf{x})$ in (7.3) is a constant times the unit matrix, so that (7.3) can be written in the form

$$\sum_{\nu=1}^{M} \left[-\alpha(\mathbf{x})\nabla^2 - \lambda_\kappa \right] |\Phi_\nu\rangle C_{\nu,\kappa} = 0 \qquad (7.23)$$

so that the secular equations become

$$\sum_{\nu=1}^{M} \langle\Phi_{\nu'}| \left[-\alpha(\mathbf{x})\nabla^2 - \lambda_\kappa \right] |\Phi_\nu\rangle C_{\nu,\kappa} = 0 \qquad (7.24)$$

In order to avoid generating non-Hermitian matrices and complex roots, we divide this equation by $\alpha(\mathbf{x})\lambda_\kappa$ so that it becomes

$$\sum_{\nu=1}^{M} \langle\Phi_{\nu'}| \left[-\frac{1}{\lambda_\kappa}\nabla^2 - \frac{1}{\alpha(\mathbf{x})} \right] |\Phi_\nu\rangle C_{\nu,\kappa} = 0 \qquad (7.25)$$

When we let the Laplacian operator act on the basis function to the right and rearrange the terms, we obtain

$$\sum_{\nu=1}^{M} \left[\langle\Phi_{\nu'}|\frac{1}{\alpha(\mathbf{x})}|\Phi_\nu\rangle - \frac{1}{\lambda_\kappa}\frac{|\mathbf{n}|^2\pi^2}{L^2}\delta_{\nu',\nu} \right] C_{\nu,\kappa} = 0 \qquad (7.26)$$

where

$$|\mathbf{n}|^2 = n_1^2 + n_2^2 + n_3^2 \qquad (7.27)$$

While avoiding non-Hermitian matrixes and complex roots, equation (7.26) is still not in the form that was discussed in Chapter 1. However, we can bring it into the desired form by changing the normalization of our basis functions, just as we did in Chapter 6.

$$|\tilde{\Phi}_\nu\rangle \equiv \frac{L}{\pi|\mathbf{n}|}|\Phi_\nu\rangle \qquad (7.28)$$

then (7.26) can be written as

$$\sum_{\nu=1}^{M} \left[\langle\tilde{\Phi}_{\nu'}|\tilde{T}|\tilde{\Phi}_\nu\rangle - \tilde{\lambda}_\kappa\delta_{\nu',\nu} \right] C_{\nu\kappa} = 0 \qquad (7.29)$$

where

$$\tilde{T} \equiv \frac{1}{\alpha(\mathbf{x})} \qquad \tilde{\lambda}_\kappa \equiv \frac{1}{\lambda_\kappa}$$

$$\langle\tilde{\Phi}_{\nu'}|\tilde{T}|\tilde{\Phi}_\nu\rangle \equiv \frac{L^2}{\pi^2|\mathbf{n'}||\mathbf{n}|}\langle\Phi_{\nu'}|\frac{1}{\alpha(\mathbf{x})}|\Phi_\nu\rangle \qquad (7.30)$$

Now suppose that inside the cube, the configuration of the inhomogeneous medium is invariant under the operations of some point group $\mathcal{G}$. The operations of $\mathcal{G}$ will not mix basis functions corresponding to different values of $|\mathbf{n}|^2$, and therefore the basis functions corresponding to a particular value of $|\mathbf{n}|^2$ can be taken as an invariant subset of the large M-dimensional basis. Solution of the secular equation, (7.26), we will automatically generate symmetry-adapted basis functions which will facilitate the solution of the large $M \times M$ problem.

As an example, let us consider the case where $\alpha(\mathbf{x})$ has the form:

$$
\alpha(\mathbf{x}) = \begin{cases} \alpha_1 & 0 < x_j < p \qquad j = 1, 2, 3 \\ \\ \alpha_2 & \text{otherwise} \end{cases} \tag{7.31}
$$

as is illustrated in Figure 7.2. Then

$$
\left\langle \Phi_{\nu'} \left| \frac{1}{\alpha(\mathbf{x})} \right| \Phi_{\nu} \right\rangle = \left(\frac{2}{L} \right)^3 \prod_{j=1}^{3} \int_0^L dx_j \sin \left(\frac{\pi n'_j x_j}{L} \right) \frac{1}{\alpha(\mathbf{x})} \sin \left(\frac{\pi n_j x_j}{L} \right)
$$

$$
= \frac{1}{\alpha_2} \delta_{\nu', \nu} + \left(\frac{1}{\alpha_1} - \frac{1}{\alpha_2} \right) \left(\frac{2}{L} \right)^3 \prod_{j=1}^{3} \int_0^p dx_j \sin \left(\frac{\pi n'_j x_j}{L} \right) \sin \left(\frac{\pi n_j x_j}{L} \right)
$$

$$
\tag{7.32}
$$

By diagonalizing the matrix $\langle \tilde{\Phi}_{\nu'} | \tilde{T} | \tilde{\Phi}_{\nu} \rangle$ for invariant blocks we can generate symmetry-adapted basis functions, as is illustrated in Figures 7.3-7.6. These can be used to find the solutions $|\Psi_\kappa(\mathbf{x}, 0)\rangle$, as shown in Figures 7.7-7.9.

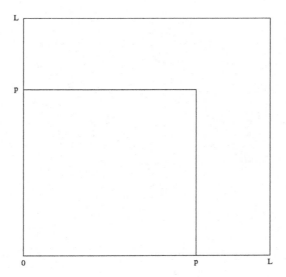

Fig. 7.2 *This diagram shows schematically the arrangement of regions in our example. A small cube of side p is situated in the corner of a large cube with side L.*

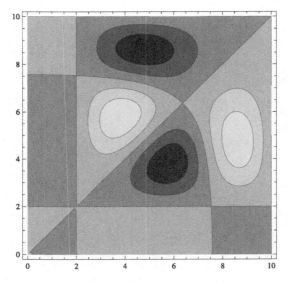

Fig. 7.3 *A symmetry-adapted basis function found by diagonalizing the invariant block corresponding to $|\mathbf{n}|^2 = 1^2 + 2^2 + 4^2 = 21$. The 3-dimensional function is shown here in the plane $x_3 = 2$. In this example, we have chosen $L = 10$, $\alpha_1 = 0.1$ and $\alpha_2 = 1$, and these parameters are the same in all of the figures that follow. This symmetry-adapted basis function corresponds to the irreducible representation A_{1u} of the group O_h.*

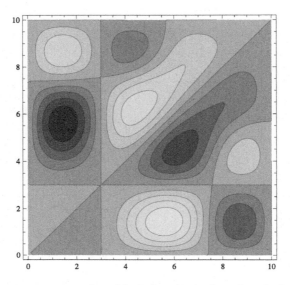

Fig. 7.4   *The same symmetry-adapted basis function is shown here in the plane $x_3 = 3$.*

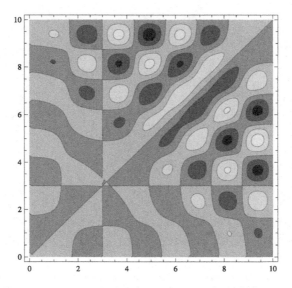

Fig. 7.5   *Another 3-dimensional A_{1u} symmetry-adapted basis function, this time found by diagonalizing the invariant block corresponding to $|\mathbf{n}|^2 = 7^2 + 8^2 + 9^2 = 194$. The function is shown in the plane $x_3 = 3$.*

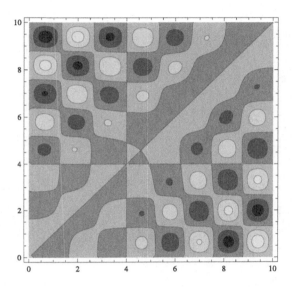

Fig. 7.6   *This figure shows the same function in the plane $x_3 = 4$.*

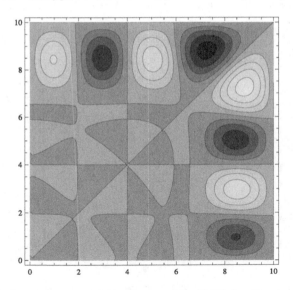

Fig. 7.7   *This figure shows a solution to equation (7.29) for the example shown in (7.31), with $p = 0.7L$, $\alpha_2 = 1$, $\alpha_1 = 0.1$ and $L = 10$. A solution, $\Psi_\kappa(\mathbf{x}, 0)$ is shown in the plane $x_3 = 0.4L = 4$. This solution to the partial differential equation (7.1) penetrates very little into the small cubic region where $x_j < 7$, $j = 1, 2, 3$. It is built up from A_{1u} symmetry-adapted basis functions of the type shown in Figures 7.3-7.6.*

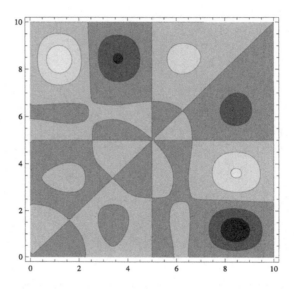

Fig. 7.8   *The same solution,* $\Psi_\kappa(\mathbf{x}, 0)$, *is shown here in the plane* $x_3 = 0.5L = 5$.

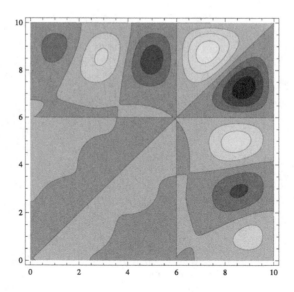

Fig. 7.9   *The same solution, but in the plane* $x_3 = 0.6L = 6$.

Chapter 8

SYMMETRY-ADAPTED SOLUTIONS BY ITERATION

8.1 Conservation of symmetry under Fourier transformation

Let us consider a d-dimensional plane wave defined by

$$e^{i\mathbf{p} \cdot \mathbf{x}} \equiv e^{i(p_1 x_1 + \cdots + p_d x_d)} \tag{8.1}$$

Expanding the wave as a Taylor series gives

$$e^{i\mathbf{p} \cdot \mathbf{x}} = \sum_{n=0}^{\infty} \frac{(i\mathbf{p} \cdot \mathbf{x})^n}{n!} = \sum_{n=0}^{\infty} \frac{(ipr)^n (\hat{\mathbf{p}} \cdot \hat{\mathbf{x}})^n}{n!} \tag{8.2}$$

where

$$\hat{\mathbf{p}} \equiv \frac{1}{p}(p_1, p_2, \ldots, p_d) \qquad p \equiv |\mathbf{p}| \tag{8.3}$$

and

$$\hat{\mathbf{x}} \equiv \frac{1}{r}(x_1, x_2, \ldots, x_d) \qquad r \equiv |\mathbf{x}| \tag{8.4}$$

Now suppose that $\mathcal{G}$ is a group of operations that leave the hyperradius invariant, and suppose that P_m^α is a group-theoretical projection operator corresponding to the mth basis function of the αth irreducible representation of $\mathcal{G}$. If we let $\tilde{P}_m^\alpha$ be the corresponding projection operator acting in momentum space, then

$$P_m^\alpha \left[(\hat{\mathbf{x}} \cdot \hat{\mathbf{p}})^n \right] = \tilde{P}_m^\alpha \left[(\hat{\mathbf{p}} \cdot \hat{\mathbf{x}})^n \right] = \tilde{P}_m^\alpha \left[(\hat{\mathbf{x}} \cdot \hat{\mathbf{p}})^n \right] \tag{8.5}$$

The first equality in equation (8.5) holds because renaming the unit vectors has no effect, i.e. operations on a set of Cartesian coordinates called $x_1, x_2, \ldots, x_d$ will be the same, no matter what names are given to the coordinates. We can also call them $p_1, p_2, \ldots, p_d$. The effect of any operation, for example an inversion or rotation, will be the same regardless of the names of the Cartesian coordinates. The second equality holds because $\hat{\mathbf{u}} \cdot \hat{\mathbf{u}}_p = \hat{\mathbf{u}}_p \cdot \hat{\mathbf{u}}$. Combining equations (8.2) and (8.5) we can see that

$$\tilde{P}_m^\alpha \left[e^{-i\mathbf{p} \cdot \mathbf{x}} \right] = P_m^\alpha \left[e^{-i\mathbf{p} \cdot \mathbf{x}} \right] \tag{8.6}$$

Now let us consider a function $f(\mathbf{x})$ which transforms under the elements of $\mathcal{G}$ like the mth basis function of the αth irreducible representation of $\mathcal{G}$. Then

$$P_m^\alpha [f(\mathbf{x})] = f(\mathbf{x}) \tag{8.7}$$

Let us now calculate the d-dimensional Fourier transform of $f(\mathbf{x})$ making use of the idempotent property of projection operators:

$$\begin{aligned}
f^t(\mathbf{p}) &= \frac{1}{(2\pi)^{d/2}} \int dx \, e^{-i\mathbf{p} \cdot \mathbf{x}} f(\mathbf{x}) \\[2mm]
&= \frac{1}{(2\pi)^{d/2}} \int dx \, e^{-i\mathbf{p} \cdot \mathbf{x}} P_n^\alpha [f(\mathbf{x})] \\[2mm]
&= \frac{1}{(2\pi)^{d/2}} \int dx \, P_m^\alpha \left[e^{-i\mathbf{p} \cdot \mathbf{x}} \right] f(\mathbf{x}) \\[2mm]
&= \frac{1}{(2\pi)^{d/2}} \int dx \, \tilde{P}_m^\alpha \left[e^{-i\mathbf{p} \cdot \mathbf{x}} \right] f(\mathbf{x}) \\[2mm]
&= \tilde{P}_m^\alpha \left[f^t(\mathbf{p}) \right]
\end{aligned} \tag{8.8}$$

Thus we see that symmetry with respect to a group of operations that leave the hyperradius invariant is conserved under a d-dimensional Fourier transformation. If $f(\mathbf{x})$ transforms like the mth basis function of the αth irreducible representation of $\mathcal{G}$, as is asserted in equation (8.7), then so does $f^t(\mathbf{p})$, as is demonstrated in equation (8.8).

8.2   The operator $-\Delta + p_\kappa^2$ and its Green's function

If atomic units are used, the Schrödinger equation for a system of N electrons moving in the potential $V(\mathbf{x})$ can be written in the form

$$\left[-\Delta + p_\kappa^2 + 2V(\mathbf{x}) \right] \Psi_\kappa(\mathbf{x}) = 0 \tag{8.9}$$

where

$$p_\kappa^2 \equiv -2E_\kappa \tag{8.10}$$

We would like to show that if

$$\Delta \equiv \sum_{j=1}^{d} \frac{\partial^2}{\partial x_j^2} \tag{8.11}$$

with $d = 3N$, then the Green's function of the operator $-\Delta + p_\kappa^2$ is given by

$$G(\mathbf{x} - \mathbf{x}') \equiv \frac{1}{(2\pi)^d} \int dp \, \frac{e^{i\mathbf{p}\cdot(\mathbf{x}-\mathbf{x}')}}{p_\kappa^2 + p^2} \tag{8.12}$$

where $e^{i\mathbf{p}\cdot(\mathbf{x}-\mathbf{x}')}$ is a d-dimensional plane wave. This Green's function will allow us to convert the Schrödinger equation to an integral form, which can be iterated. Applying the operator $-\Delta + p_\kappa^2$ to both sides of equation (8.12), we obtain

$$\left[-\Delta + p_\kappa^2\right] G(\mathbf{x} - \mathbf{x}') = \frac{1}{(2\pi)^d} \int dp \, e^{i\mathbf{p}\cdot(\mathbf{x}-\mathbf{x}')} = \delta(\mathbf{x} - \mathbf{x}') \tag{8.13}$$

Thus $G(\mathbf{x} - \mathbf{x}')$, as defined in (8.12), is seen to be the Green's function of $-\Delta + p_\kappa^2$.

8.3 Conservation of symmetry under iteration of the Schrödinger equation

We will now show that

$$\Psi_\kappa(\mathbf{x}) = -2 \int dx' \, G(\mathbf{x} - \mathbf{x}') V(\mathbf{x}') \Psi_\kappa(\mathbf{x}') \tag{8.14}$$

is an integral form of the Schrödinger equation for a system of particles with equal masses. (We assume here that we are dealing with an N-electron system, so that in atomic units, the masses are all equal to 1.) We will show below that it is possible to iterate the Schrödinger equation when it is written in the form shown in equation (8.14).

If we act on both sides of (8.14) with $-\Delta + p_\kappa^2$, then with the help of (8.13), we obtain:

$$\left[-\Delta + p_\kappa^2\right] \Psi_\kappa(\mathbf{x}) = -2 \left[-\Delta + p_\kappa^2\right] \int dx' \, G(\mathbf{x} - \mathbf{x}') V(\mathbf{x}') \Psi_\kappa(\mathbf{x}')$$

$$= -2 \int dx' \, \delta(\mathbf{x} - \mathbf{x}') V(\mathbf{x}') \Psi_\kappa(\mathbf{x}')$$

$$= -2V(\mathbf{x}) \Psi_\kappa(\mathbf{x}) \tag{8.15}$$

Thus if $G(\mathbf{x} - \mathbf{x}')$ is defined by (8.12), then (8.14) is an integral form of the N-electron Schrödinger equation (8.9).

Let us now consider a group of operations $\mathcal{G}$ that not only leave the hyperradius invariant but also commute with the potential $V(\mathbf{x})$. Iteration of the integral form of the N-particle Schrödinger equation (8.14) will then preserve the symmetry of an ith-order trial function, $\Psi_\kappa^{(i)}(\mathbf{x})$. The iteration involves a double Fourier transformation, but since symmetry is conserved under both these Fourier transformations, and since $(p_\kappa^2 + p^2)^{-1}$ is invariant, the symmetry of the trial function is conserved under a Green's function iteration of the form

$$\Psi_\kappa^{(i+1)}(\mathbf{x}) = -2 \int dx' \, G(\mathbf{x} - \mathbf{x}')V(\mathbf{x}')\Psi_\kappa^{(i)}(\mathbf{x}') \tag{8.16}$$

If

$$P_m^\alpha \left[V(\mathbf{x})\Psi_\kappa^{(i)}(\mathbf{x}) \right] = V(\mathbf{x})P_m^\alpha \left[\Psi_\kappa^{(i)}(\mathbf{x}) \right] = V(\mathbf{x})\Psi_\kappa^{(i)}(\mathbf{x}) \tag{8.17}$$

then, because of conservation of symmetry under Fourier transformation,

$$\tilde{P}_m^\alpha \left[\left(V\Psi_\kappa^{(i)} \right)^t (\mathbf{p}) \right] = \left(V\Psi_\kappa^{(i)} \right)^t (\mathbf{p}) \tag{8.18}$$

In other words, both $V(\mathbf{x})\Psi_\kappa^{(i)}(\mathbf{x})$ and $\left(V\Psi_\kappa^{(i)} \right)^t (\mathbf{p})$ transform like the mth basis function of the αth irreducible representation of $\mathcal{G}$. From equations (8.14), (8.12) and (8.16) we have

$$\Psi_\kappa^{(i+1)}(\mathbf{x}) = -2 \int dx' \, G(\mathbf{x} - \mathbf{x}')V(\mathbf{x}')\Psi_\kappa^{(i)}(\mathbf{x}')$$

$$= -\frac{2}{(2\pi)^{d/2}} \int dp \, \frac{e^{i\mathbf{p}\cdot\mathbf{x}}}{p_\kappa^2 + p^2} \left(V\Psi_\kappa^{(i)} \right)^t (\mathbf{p}) \tag{8.19}$$

Applying our projection operator to the iterated solution we have

$$P_m^\alpha \left[\Psi_\kappa^{i+1}(\mathbf{x}) \right] = -\frac{2}{(2\pi)^{d/2}} \int dp \, \frac{P_m^\alpha \left[e^{i\mathbf{p}\cdot\mathbf{x}} \right]}{p_\kappa^2 + p^2} \left(V\Psi_\kappa^{(i)} \right)^t (\mathbf{p})$$

$$= -\frac{2}{(2\pi)^{d/2}} \int dp \, \frac{\tilde{P}_m^\alpha \left[e^{i\mathbf{p}\cdot\mathbf{x}} \right]}{p_\kappa^2 + p^2} \left(V\Psi_\kappa^{(i)} \right)^t (\mathbf{p})$$

$$= -\frac{2}{(2\pi)^{d/2}} \int dp \, \frac{e^{i\mathbf{p}\cdot\mathbf{x}}}{p_\kappa^2 + p^2} \tilde{P}_m^\alpha \left[\left(V\Psi_\kappa^{(i)} \right)^t (\mathbf{p}) \right]$$

$$= -\frac{2}{(2\pi)^{d/2}} \int dp \, \frac{e^{i\mathbf{p}\cdot\mathbf{x}}}{p_\kappa^2 + p^2} \left(V\Psi_\kappa^{(i)} \right)^t (\mathbf{p})$$

$$= \Psi_\kappa^{i+1}(\mathbf{x}) \tag{8.20}$$

Thus we see that if the operations of $\mathcal{G}$ commute with $V(\mathbf{x})$, Green's function iteration of the N-electron Schrödinger equation preserves symmetry under the operations of $\mathcal{G}$.

8.4   Evaluation of the integrals

The iteration process can be facilitated by using the generalized Sturmian expansion of a plane wave, which is discussed in Appendix B.

$$e^{-i\mathbf{p}\cdot\mathbf{x}'} = (2\pi)^{d/2} \left(\frac{p_\kappa^2 + p^2}{2p_\kappa^2}\right) \sum_\nu \Phi_\nu^t(\mathbf{p})\Phi_\nu^*(\mathbf{x}') \qquad (8.21)$$

where the functions $\Phi_\nu(\mathbf{x}')$ are a set of generalized Sturmian basis functions. This expansion is not pointwise convergent, but holds in the sense of distributions. (When we say that a relationship is valid in the sense of distributions we mean that if the left-hand side is placed inside an integral, multiplied by some function, and integrated over the relevant variables, the result is the same as it is when the right-hand side is treated the same way.) We can substitute this expansion into (8.12) and (8.14) we obtain the relationship

$$\begin{aligned}
\Psi_\kappa(\mathbf{x}) &= -2 \int dx' \; G(\mathbf{x} - \mathbf{x}')V(\mathbf{x}')\Psi_\kappa(\mathbf{x}') \\
&= -\frac{2}{(2\pi)^d} \int dp \; \frac{e^{i\mathbf{p}\cdot\mathbf{x}}}{p_\kappa^2 + p^2} \int dx' \; e^{-i\mathbf{p}\cdot\mathbf{x}'} V(\mathbf{x}')\Psi_\kappa(\mathbf{x}') \\
&= -\frac{2}{2p_\kappa^2} \sum_\nu \frac{1}{(2\pi)^{d/2}} \int dp \; e^{i\mathbf{p}\cdot\mathbf{x}}\Phi_\nu^t(\mathbf{p}) \int dx' \; \Phi_\nu^*(\mathbf{x}')V(\mathbf{x}')\Psi_\kappa(\mathbf{x}') \\
&= -\frac{2}{2p_\kappa^2} \sum_\nu \Phi_\nu(\mathbf{x}) \int dx' \; \Phi_\nu^*(\mathbf{x}')V(\mathbf{x}')\Psi_\kappa(\mathbf{x}') \qquad (8.22)
\end{aligned}$$

If we compare the first and last lines of equation (8.22), we can make the identification (in the sense of distributions)

$$G(\mathbf{x} - \mathbf{x}') = \frac{1}{2p_\kappa^2} \sum_\nu \Phi_\nu(\mathbf{x})\Phi_\nu^*(\mathbf{x}') \qquad (8.23)$$

In other words, for any complete set of generalized Sturmian basis functions, we can make the identification

$$\frac{1}{2p_\kappa^2} \sum_\nu \Phi_\nu(\mathbf{x})\Phi_\nu^*(\mathbf{x}') = \frac{1}{(2\pi)^d} \int dp \; \frac{e^{i\mathbf{p}\cdot(\mathbf{x}-\mathbf{x}')}}{p_\kappa^2 + p^2} \qquad (8.24)$$

which is valid in the sense of distributions.

Suppose that we now wish to iterate the Schrödinger equation in the form shown in equation (8.22). Substituting an initial solution into the integral on the right-hand side, we obtain a first-iterated solution. This can in turn be substituted into the integral on the right-hand side, yielding a second-iterated solution, and so on.

$$\Psi_\kappa^{(1)}(\mathbf{x}) = -\frac{1}{p_\kappa^2} \sum_\nu \Phi_\nu(\mathbf{x}) \int dx' \ \Phi_\nu^*(\mathbf{x}')V(\mathbf{x}')\Psi_\kappa^{(0)}(\mathbf{x}')$$

$$\Psi_\kappa^{(2)}(\mathbf{x}) = -\frac{1}{p_\kappa^2} \sum_\nu \Phi_\nu(\mathbf{x}) \int dx' \ \Phi_\nu^*(\mathbf{x}')V(\mathbf{x}')\Psi_\kappa^{(1)}(\mathbf{x}')$$

$$\vdots \quad \vdots \qquad \qquad \vdots \qquad\qquad\qquad\qquad (8.25)$$

If we now let

$$T_{\nu',\nu} \equiv -\frac{1}{p_\kappa} \int dx' \ \Phi_{\nu'}^*(\mathbf{x}')V(\mathbf{x}')\Phi_\nu(\mathbf{x}') \qquad (8.26)$$

and

$$\Psi_\kappa^{(i)}(\mathbf{x}') = \sum_\nu \Phi_\nu(\mathbf{x}')B_{\nu,\kappa}^{(i)} \qquad (8.27)$$

then we obtain the relationship

$$\Psi^{(i+1)}(\mathbf{x}) = -\frac{1}{p_\kappa^2} \sum_{\nu',\nu} \Phi_{\nu'}(\mathbf{x}) \int dx' \ \Phi_{\nu'}^*(\mathbf{x}')V(\mathbf{x}')\Phi_\nu(\mathbf{x}')B_{\nu,\kappa}^{(i)}$$

$$= \frac{1}{p_\kappa} \sum_{\nu',\nu} \Phi_{\nu'}(\mathbf{x})T_{\nu',\nu}B_{\nu,\kappa}^{(i)}$$

$$= \sum_{\nu'} \Phi_{\nu'}(\mathbf{x})B_{\nu',\kappa}^{(i+1)} \qquad (8.28)$$

where

$$B_{\nu',\kappa}^{(i+1)} = \frac{1}{p_\kappa} \sum_\nu T_{\nu',\nu}B_{\nu,\kappa}^{(i)} \qquad (8.29)$$

Thus the iteration can easily be performed in practice provided that we are able to evaluate the matrix elements $T_{\nu',\nu}$. Equations (8.27)-(8.29) can be used to iterate the N-electron Schrödinger equation even when the generalized Sturmian basis set $\Phi_\nu(\mathbf{x})$ is not complete, but of course the iterated solutions will never leave the part of Sobolev space spanned by the basis set.

8.5 Generation of symmetry-adapted basis functions by iteration

An initial solution can be obtained by solving the Sturmian secular equations

$$\sum_{\nu \in W_0} [T_{\nu',\nu} - p_\kappa \delta_{\nu',\nu}] B_{\nu,\kappa}^{(0)} = 0 \qquad \nu' \in W_0 \qquad (8.30)$$

with a truncated basis set contained in an invariant subset W_0. Equation (8.29) can then be used as a criterion for automatic selection of a larger basis set to be used in a more accurate version of the secular equations. Let

$$B_{\nu',\kappa}^{(1),\alpha,n} = \frac{1}{p_\kappa} \sum_{\nu \in W_0} T_{\nu',\nu} B_{\nu,\kappa}^{(0),\alpha,n} \qquad \nu' \in W_1 \qquad (8.31)$$

The superscript symmetry indices α, n indicate that an eigenvector of a particular symmetry has been chosen from among the solutions to equation (8.30). If the initial solution has a symmetry corresponding to one of the irreducible representations of the symmetry group of $V(\mathbf{x})$, the first-iterated solution will be of the same symmetry, as was discussed above. Let the target function be

$$|\eta_\kappa^{(0)\,\alpha,n}\rangle = \sum_\nu |\Phi_\nu\rangle B_{\nu,\kappa}^{(0)\,\alpha,n} \qquad \nu \in W_0 \qquad (8.32)$$

Then for $\nu' \in W_1$,

$$\frac{1}{p_\kappa} \sum_{\nu'} |\Phi_{\nu'}\rangle \left\langle \Phi_{\nu'} \Big| T \Big| \eta_\kappa^{(0)\,\alpha,n} \right\rangle = \frac{1}{p_\kappa} \sum_{\nu'} \sum_\nu |\Phi_{\nu'}\rangle \langle \Phi_{\nu'}|T|\Phi_\nu\rangle B_{\nu,\kappa}^{(0)\,\alpha,n}$$

$$= \frac{1}{p_\kappa} \sum_{\nu'} \sum_\nu |\Phi_{\nu'}\rangle T_{\nu',\nu} B_{\nu,\kappa}^{(0)\,\alpha,n}$$

$$= \sum_{\nu'} |\Phi_{\nu'}\rangle B_{\nu',\kappa}^{(1)\,\alpha,n} = |\eta_\kappa^{(1)\,\alpha,n}\rangle \qquad (8.33)$$

Thus, by performing the operations shown in equation (8.33), we obtain a new symmetry-adapted basis function with the same symmetry as the target function $|\eta_\kappa^{(0)\,\alpha,n}\rangle$. There is a difficulty, however. The difficulty is that some of the flexibility of our basis set may be lost by lumping several linearly independent symmetry-adapted basis functions together in a single function. We must do a little additional work find the smallest possible combinations of our primitive basis functions $|\Phi_\nu\rangle$ that still have the correct symmetry.

To see how the simplest possible symmetry-adapted basis functions may be obtained from a target function by iteration, let us consider a calculation of atomic states using Goscinskian configurations. Then the coefficients $B_{\nu',\kappa}^{(1),\alpha,n}$ will depend on the nuclear charge. If we now construct a vector whose components are

$$B_{\nu',\kappa}^{(1),\alpha,n}(Z_1)/B_{\nu',\kappa}^{(1),\alpha,n}(Z_2) \qquad \nu' \in W_1 \qquad (8.34)$$

for the non-zero values of the coefficients, the simplest possible symmetry-adapted basis functions can be found by collecting coefficients corresponding to particular values of the ratio.

8.6   A simple example

As a simple example of iteration, we might consider the Generalized Sturmian Method applied to atoms and atomic ions, as was discussed in Chapter 4. In that case, the generalized Sturmian basis set will consist of Goscinskian configurations of the type shown in equations (4.1)-(4.16). To make the example more specific, let us consider the ^{4}P states of the lithiumlike isoelectronic series. A "target function" with this symmetry can be obtained by diagonalizing the (28×28)-dimensional block of matrix elements $T_{\nu',\nu}$ corresponding to the $\mathcal{R}_\nu$ value

$$\mathcal{R}_\nu = \sqrt{\frac{1}{1^2} + \frac{1}{2^2} + \frac{1}{2^2}} = \sqrt{\frac{3}{2}} \qquad (8.35)$$

This is the block of possible configurations with 1 electron in the $n = 1$ shell and 2 electrons in the $n = 2$ shell. Suppose that we now choose as our target function the eigenfunction corresponding to $(1s \uparrow)(2s \uparrow)(2p_1 \uparrow)$. Let us call this basis function $\Phi_1(\mathbf{x})$. The next step is to choose a much larger invariant subset of basis functions W_1 and to evaluate all the matrix elements $T_{\nu,1}$ for $\nu \in W_1$. Continuing in this way, we found a set of 462 symmetry-adapted configurations especially adapted to ^{4}P states of the lithium-like isoelectronic series. The energies (in Hartrees) of the lowest ^{4}P states found by diagonalizing the block of $T_{\nu',\nu}$ based on these configurations are shown in Table 8.1.

8.7 An alternative expansion of the Green's function that applies to the Hamiltonian formulation of physics

In the Hamiltonian formulation of physics, we solve equations of the form

$$[H_0 - E_\nu]\,\Phi_\nu(\mathbf{x}) = 0 \tag{8.36}$$

and obtain basis sets (eigenfunctions of H_0) that obey orthonormality relations of the form

$$\int dx\;\Phi_{\nu'}^*(\mathbf{x})\Phi_\nu(\mathbf{x}) = \delta_{\nu',\nu}$$

$$\int dp\;\Phi_{\nu'}^{t*}(\mathbf{p})\Phi_\nu^t(\mathbf{p}) = \delta_{\nu',\nu} \tag{8.37}$$

These more conventional orthonormality relations lead to an alternative expansion of a d-dimensional plane wave, equation (B.60).

$$e^{-i\mathbf{p}\cdot\mathbf{x}} = (2\pi)^{d/2}\sum_\nu \Phi_\nu^t(\mathbf{p})\Phi_\nu^*(\mathbf{x}) \tag{8.38}$$

If we substitute this alternative plane-wave expansion (which is appropriate for eigenfunctions of a zeroth-order Hamiltonian H_0) into equation (8.22), we obtain the relation

$$\Psi_\kappa(\mathbf{x}) = -2\sum_\nu \tilde{\Phi}_\nu(\mathbf{x})\int dx'\;\Phi_\nu^*(\mathbf{x}')V(\mathbf{x}')\Psi_\kappa(\mathbf{x}') \tag{8.39}$$

where

$$\tilde{\Phi}_\nu(\mathbf{x}) = \frac{1}{(2\pi)^{d/2}}\int dp\;\frac{e^{i\mathbf{p}\cdot\mathbf{x}}}{(p^2 + p_\kappa^2)}\Phi_\nu^t(\mathbf{p}) \tag{8.40}$$

and $p_\kappa^2 = -2E_\kappa$. Comparing this with (8.14) we obtain an alternative representation of the Green's function of the Schrödinger equation:

$$G(\mathbf{x} - \mathbf{x}') = \sum_\nu \tilde{\Phi}_\nu(\mathbf{x})\Phi_\nu^*(\mathbf{x}') \tag{8.41}$$

Table 8.1: This table shows energies of the lowest ^4P states of the lithiumlike isoelectronic series, calculated using a basis of 462 symmetry–adapted Goscinskian configurations. The appropriate symmetry–adapted basis set was found by iteration. The second
column shows the calculated energies, crudely corrected for relativistic effects using the
method described in Chapter 4. The final column shows experimental energies from the
NIST database.

	calc.	corr.	expt.
Z=3	−5.3587	−5.3592	−5.3660
Z=4	−10.053	−10.059	−10.067
Z=5	−16.250	−16.255	no data
Z=6	−23.950	−23.960	no data
Z=7	−33.151	−33.170	−33.162
Z=8	−43.852	−43.887	−43.899
Z=9	−56.054	−56.110	no data
Z=10	−69.757	−69.842	−69.869
Z=11	−84.959	−85.085	−85.107
Z=12	−101.66	−101.84	−101.86

Appendix A

REPRESENTATION THEORY OF FINITE GROUPS

A.1   Basic definitions

Definition A.1. A *group* is a set G together with a composition operation $\circ : G \times G \to G$ that obeys the following conditions:

(1) G is closed under composition: if $a, b \in G$, then $a \circ b \in G$.
(2) G contains a neutral element e such that $ae = a = ea$ for every $a \in G$.
(3) Every element $a \in G$ has a unique inverse a^{-1} with the property that $aa^{-1} = e = a^{-1}a$.
(4) Composition is associative: $a \circ (b \circ c) = (a \circ b) \circ c$.

Groups are usually written either additively ($a \circ b = a + b$) or multiplicatively ($a \circ b = ab$). For additively written groups, we write $-a$ for the inverse to a, and 0 for the neutral element e. For multiplicatively written groups, we write a^{-1} or $1/a$ for the inverse, and 1 for the neutral element. The notation has no bearing, however, on the properties of G as a group.

The number of elements of G is $|G|$ and is called the *order* of the group. If $|G| < \infty$, we say that the group is *finite*. In this book, we will use the shorthand $g = |G|$ to denote the order of the group.

Example A.1. A few simple examples of groups are: the real or complex numbers under addition, any vector space under addition, the real or complex numbers, except for zero, with multiplication as composition, the complex circle $\{e^{i\theta} | \theta \in [0; 2\pi[\}$ with multiplication, and the set of $n \times n$ invertible matrices under matrix multiplication.

Example A.2. As a slightly more complicated example, we might think of a molecule which is symmetric with respect to rotations through an angle of $2\pi/3$ about some axis but which has no other symmetry. Then the set of geometrical operations that leave the molecule invariant form a group containing 3 elements: the identity element; a rotation through an angle $2\pi/3$ about the axis of symmetry, and a rotation through an angle $4\pi/3$ about the same axis. Let us denote these operations respectively by E, C_3, and C_3^{-1}. We can easily construct a multiplication table for the group. If we do so, *each element of the group will appear once and only once in any row or column of the multiplication table*. This follows from the fact that $AX = B$ has one and only one solution among the group elements. Since A^{-1} and B belong to the group, and since the product of any two elements belongs to the group, $X = A^{-1}B$ is also a uniquely-defined element. Now suppose that the element B appears more than once in the Ath row of the multiplication table. Then $AX = B$ will have more than one solution which is impossible. Since no element can appear more than once, each element must appear once because there are g elements and g places in the row, all of which have to be filled.

A.2 Representations of geometrical symmetry groups

The elements of a geometrical symmetry group are linear coordinate transformations. Such transformations have the form

$$X^i = \sum_{j=1}^{d} \frac{\partial X^i}{\partial x^j} x^j + b^i \tag{A.1}$$

where $\partial X^i/\partial x^j$ and b^i are constants.

Now consider a set of functions Φ_1, Φ_2, $\ldots$, Φ_M. We can use equation (A.1) to express $\Phi_1(\mathbf{x})$ as a function of $\mathbf{X}$. If we then expand the resulting function of $\mathbf{X}$ in terms of the other $|\Phi_n\rangle$'s, we shall obtain a relation of the form

$$\Phi_n(\mathbf{x}) = \sum_{n'} \Phi_{n'}(\mathbf{X}) D_{n',n} \tag{A.2}$$

If we denote the coordinate transformation in equation (A.1) by the symbol G, we can rewrite equations (A.1) and (A.2) in the form:

$$\mathbf{X} = G_j \mathbf{x}$$
$$\Phi_n(\mathbf{x}) \equiv \Phi_n(G_j^{-1}\mathbf{X}) \equiv G_j\Phi_n(\mathbf{X})$$
$$= \sum_{n'} \Phi_{n'}(\mathbf{X}) D_{n',n}(G) \tag{A.3}$$

In this sense, the coordinate transformation defines an operator G_j, and $D_{n',n}(G_j)$ is a matrix representing G_j. It can easily be shown that the matrices representing a set of operators G_1, G_2,..., G_g in a given basis, obey the same multiplication table as the operators themselves. For example, if we know that

$$C_3 C_3^{-1} = E \qquad (A.4)$$

and that

$$C_3 \Phi_n = \sum_{n'} \Phi_{n'} D_{n',n}(C_3)$$

$$C_3^{-1} \Phi_n = \sum_{n'} \Phi_{n'} D_{n',n}(C_3^{-1}) \qquad (A.5)$$

$$E \Phi_n = \sum_{n'} \Phi_{n'} D_{n',n}(E)$$

then it follows that:

$$C_3 C_3^{-1} \Phi_n = \sum_{n'} C_3 \Phi_{n'} D_{n',n}(C_3^{-1})$$

$$= \sum_{n''} \Phi_{n''} \left\{ \sum_{n'} D_{n'',n'}(C_3) D_{n',n}(C_3^{-1}) \right\} \qquad (A.6)$$

$$= E \Phi_n = \sum_{n''} \Phi_{n''} D_{n'',n}(E)$$

so that we must have

$$D_{n'',n}(E) = \sum_{n'} D_{n'',n'}(C_3) D_{n',n}(C_3^{-1}) \qquad (A.7)$$

Thus *given any set of basis functions* Φ_1, Φ_2, ..., Φ_M *which mix together under the elements of a group* G_1, G_2,..., G_g, *we can obtain a set of matrices* $D_{n',n}(G_j)$ *defined by the relationships*

$$G_j \Phi_n = \sum_{n'} \Phi_{n'} D_{n',n}(G_j) \qquad j = 1, 2, \ldots, g \qquad (A.8)$$

These matrices will obey the same multiplication table as the operators G_1, G_2,..., G_g, *and they are said to form a matrix representation of the group.*

A.3 Similarity transformations

Now let us consider another representation, $D'_{m',m}(G_j)$, based on a set of functions Φ'_1, Φ'_2, ..., Φ'_M which are related to our original set Φ_1, Φ_2, ..., Φ_M by the transformation:

$$\Phi'_m = \sum_n \Phi_n S_{n,m}$$

$$\Phi_n = \sum_m \Phi'_m S_{m,n}^{-1} \qquad (A.9)$$

The primed representation is defined by the relationship

$$G_j \Phi'_m = \sum_{m'} \Phi'_{m'} D'_{m',m}(G_j) \qquad j = 1, 2, \ldots, g \qquad (A.10)$$

Then from equations (A.8)-(A.10) we have

$$
\begin{aligned}
G_j \Phi'_m &= \sum_{m'} \Phi'_{m'} D'_{m',m}(G_j) \\
&= G_j \sum_{n} \Phi_n S_{n,m} \\
&= \sum_{n,n'} \Phi_{n'} D_{n',n}(G_j) S_{n,m} \\
&= \sum_{m',n,n'} \Phi'_{m'} S^{-1}_{m',n'} D_{n',n}(G_j) S_{n,m}
\end{aligned}
\qquad (A.11)
$$

so that we must have

$$D'_{m',m}(G_j) = \sum_{n,n'} S^{-1}_{m',n'} D_{n',n}(G_j) S_{n,m} \qquad (A.12)$$

or

$$D' = S^{-1} D S \qquad (A.13)$$

A transformation of this type, where the matrix S need not be unitary, is called a 'similarity transformation'.

A.4 Characters and reducibility

The character $\chi(G_j)$ of the matrix $D_{n',n}(G_j)$ is defined as the sum of the diagonal elements:

$$\chi(G_j) \equiv \sum_{n} D_{n,n}(G_j) \qquad (A.14)$$

We would like to show that the character of each element in a representation of a finite group is invariant under a similarity transformation. From equations (A.12) and (A.14) we have:

$$\chi'(G_j) \equiv \sum_m D'_{m,m}(G_j)$$

$$= \sum_{m,n,n'} S^{-1}_{m,n'} D_{n',n}(G_j) S_{n,m}$$

$$= \sum_{n,n'} \left(\sum_m S_{n,m} S^{-1}_{m,n'} \right) D_{n',n}(G_j) \qquad \text{(A.15)}$$

$$= \sum_{n,n'} \delta_{n',n} D_{n',n}(G_j)$$

$$= \sum_n D_{n,n}(G_j) = \chi(G_j) \qquad \text{q.e.d.}$$

If two representations are connected by a similarity transformation, then they are said to be 'equivalent'. From (A.15) it follows that when two representations are equivalent, then $\chi'(G_j) = \chi(G_j)$ for $j = 1, 2, \ldots, g$.

Sometimes it is possible by means of a similarity transformation to bring all the elements of a representation into a block-diagonal form. In other words it may be possible to bring $D'_{m',m}(G_j)$ into a form where the non-zero elements are confined blocks along the diagonal, the blocks being the same for all the group elements. To express the same idea differently, it is sometimes possible to go over by means of a similarity transformation from the original basis set, Φ_1, Φ_2, ..., Φ_M to a new basis set Φ'_1, Φ'_2, ..., Φ'_M which can be divided into two or more subsets, each of which mixes only with itself under the operations $G_1, G_2, \ldots, G_g$. *A representation based on two or more subsets of basis functions which mix only with themselves under the operations of the group is said to be 'reduced'. Whenever it is possible to bring a representation into a reduced form by means of a similarity transformation, it is said to be 'reducible'. Whenever this is not possible, the representation is said to be 'irreducible'.*

Table A.1 Multiplication table for the group C_3

	E	C_3	C_3^{-1}
E	E	C_3	C_3^{-1}
C_3	C_3	C_3^{-1}	E
C_3^{-1}	C_3^{-1}	E	C_3

Table A.2 Character table for the group C_3

	E	C_3	C_3^{-1}
A	1	1	1
Γ_c	1	$e^{i(2\pi/3)}$	$e^{-i(2\pi/3)}$
Γ_c^*	1	$e^{-i(2\pi/3)}$	$e^{i(2\pi/3)}$

A.5 The great orthogonality theorem

A unitary matrix is a matrix whose conjugate transpose (Hermitian adjoint) is equal to its inverse. It is always possible, by means of a similarity transformation, to bring the matrix representations of a finite group into unitary form. Now let $D^\alpha_{n',n}(G_j)$ and $D^\beta_{m',m}(G_j)$ be two unitary irreducible representations of a finite group of order g. The great orthogonality theorem, from which much of the power of representation theory is derived, states that

$$\frac{d_\alpha}{g} \sum_{j=1}^{g} D^{\alpha*}_{n',n}(G_j) D^\beta_{m',m}(G_j) = \delta_{\alpha,\beta} \delta_{n',m'} \delta_{n,m} \tag{A.16}$$

where d_α is the dimension of the matrices $D^\alpha_{n',n}(G_j)$.

The proof of the great orthogonality theorem depends on Schur's lemma, which states that if A is a matrix that commutes with every matrix $D^\alpha_{n',n}(G_j)$, $j = 1, 2, \ldots, g$ in a unitary irreducible representation of a finite group, then A must be either zero or a multiple of the unit matrix.

The proof of Schur's lemma is as follows: If A commutes with $D^\alpha_{n',n}(G_j)$, $j = 1, 2, \ldots, g$, then so does its conjugate transpose $A^\dagger$. Therefore we can let A be Hermitian without loss of generality, and we can diagonalize A by means of a unitary transformation:

$$UAU^{-1} = A^{(d)} \tag{A.17}$$

where $A^{(d)}$ is diagonal. Then

$$U^{-1}A^{(d)}UD(G_j) - D(G_j)U^{-1}A^{(d)}U = 0, \qquad j = 1, 2, \ldots, g \tag{A.18}$$

Multiplying on the left by U and on the right by U^{-1} then yields

$$A^{(d)}UD(G_j)U^{-1} - UD(G_j)U^{-1}A^{(d)} = 0, \qquad j = 1, 2, \ldots, g \tag{A.19}$$

Thus we can write

$$A^{(d)}D'(G_j) - D'(G_j)A^{(d)} = 0, \qquad j = 1, 2, \ldots, g \tag{A.20}$$

where

$$D'(G_j) \equiv UD(G_j)U^{-1} \tag{A.21}$$

Since $A^{(d)}$ is diagonal we can write $A^{(d)}_{n',n} = A^{(d)}_n \delta_{n',n}$. Thus with the indices written out, (A.20) becomes:

$$\sum_{n'} \left(A^{(d)}_{n'} \delta_{n'',n'} D'^\alpha_{n',n}(G_j) - D'^\alpha_{n'',n'}(G_j) A^{(d)}_n \delta_{n',n} \right) = 0, \qquad j = 1, \ldots, g$$

from which it follows that

$$\left(A^{(d)}_{n'} - A^{(d)}_n \right) D'^\alpha_{n',n}(G_j) = 0, \qquad j = 1, 2, \ldots, g \tag{A.22}$$

Without loss of generality, we can choose U in such a way that repeated eigenvalues of $A^{(d)}$ are grouped together along the diagonal. Then $A_{n''}^{(d)} \neq A_n^{(d)}$ implies that

$$D_{n',n}'^{\alpha}(G_j) = D_{n',n}'^{\dagger\alpha}(G_j) = D_{n,n'}'^{\alpha}(G_j^{-1}) = 0, \quad j = 1, 2, \ldots, g \qquad (A.23)$$

Thus $D_{n',n}'^{\alpha}(G_j)$ can only have non-zero elements in the blocks that correspond to repeated eigenvalues of $A^{(d)}$ and it would therefore be reducible unless all of the eigenvalues are equal, which would contradict the original assumption of irreducibility. This proves Schur's lemma.

Having demonstrated the validity of Schur's lemma, we are now in a position to prove the great orthogonality relation. To do so we define the matrix M by the relationship

$$M \equiv \sum_{j=1}^{g} D^{\alpha}(G_j) X D^{\beta}(G_j^{-1}) \qquad (A.24)$$

where X is an arbitrary matrix of appropriate dimensions to make matrix multiplication possible and where $D^{\alpha}(G_j)$ and $D^{\beta}(G_j)$ are unitary irreducible representations of the finite group. Then

$$D^{\alpha}(G_i) M D^{\beta}(G_i^{-1}) = \sum_{j=1}^{g} D^{\alpha}(G_i) D^{\alpha}(G_j) X D^{\beta}(G_j^{-1}) D^{\beta}(G_i^{-1})$$

$$= \sum_{k=1}^{g} D^{\alpha}(G_k) X D^{\beta}(G_k^{-1}) = M \qquad (A.25)$$

where $G_i G_j = G_k$ and where we have used the fact that each group element appears once and only once in every row of the multiplication table to replace the sum over j by a sum over k. Multiplying (A.25) from the right by $D^{\beta}(G_i)$ we obtain:

$$D^{\alpha}(G_i) M = M D^{\beta}(G_i) \qquad i = 1, 2, \ldots, g \qquad (A.26)$$

Then, according to Schur's lemma, M must be either zero or a multiple of the unit matrix. Multiplying (A.26) from the left by M^{-1} we obtain:

$$M^{-1} D^{\alpha}(G_i) M = D^{\beta}(G_i) \qquad i = 1, 2, \ldots, g \qquad (A.27)$$

from which we can see that if M is not the null matrix, then the irreducible representations $D^{\alpha}(G_i)$ and $D^{\beta}(G_i)$ must be the same, i.e., if M is not the null matrix, $\alpha = \beta$.

Case 1

Let us first consider the case where M is the null matrix and where $\alpha \neq \beta$. Then putting indices into (A.24) we have:

$$\sum_{j=1}^{g}\sum_{n=1}^{d_\alpha}\sum_{m'=1}^{d_\beta} D^\alpha_{n',n}(G_j)X_{n,m'}D^\beta_{m',m}(G_j^{-1}) = 0 \tag{A.28}$$

But $X_{n,m'}$ is arbitrary, and therefore (A.28) can only hold for all cases if

$$\sum_{j=1}^{g} D^\alpha_{n',n}(G_j)D^\beta_{m',m}(G_j^{-1}) = 0, \qquad \alpha \neq \beta \tag{A.29}$$

Case 2

Now let us consider the second possibility: Suppose that $\alpha = \beta$ and $M = cI$, where I is the $d_\alpha \times d_\alpha$ dimensional identity matrix and where c is a constant. Then

$$M = cI = \sum_{j=1}^{g} D^\alpha(G_j)XD^\alpha(G_j^{-1}) \tag{A.30}$$

Taking the trace on both sides of (A.30) yields

$$\mathrm{tr}[M] = c\,\mathrm{tr}[I] = c\,d_\alpha = \sum_{j=1}^{g}\mathrm{tr}\left[D^\alpha(G_j)XD^\alpha(G_j^{-1})\right] = g\,\mathrm{tr}[X]$$

since the trace of a matrix is invariant under a unitary transformation. From (A.31) we have

$$c = \frac{g}{d_\alpha}\,\mathrm{tr}[X] \tag{A.31}$$

Putting indices into (A.30) we have

$$cI = I\frac{g}{d_\alpha}\,\mathrm{tr}[X] = \sum_{j=1}^{g}\sum_{r=1}^{d_\alpha}\sum_{s=1}^{d_\alpha} D^\alpha_{n',r}(G_j)X_{r,s}D^\alpha_{s,m'}(G_j^{-1}) \tag{A.32}$$

where I is the identity matrix. Because X is arbitrary, we can let

$$X_{r,s} = \delta_{r,n}\delta_{s,m} \tag{A.33}$$

Then

$$\mathrm{tr}[X] = \delta_{n,m} \tag{A.34}$$

and (A.32) becomes

$$\sum_{j=1}^{g} D^\alpha_{n',n}(G_j)D^\alpha_{m,m'}(G_j^{-1}) = \frac{g}{d_\alpha}\,\delta_{n,m}\delta_{n',m'} \tag{A.35}$$

Making use of the unitarity of the representations we have

$$\sum_{j=1}^{g} D_{n',n}^{\alpha}(G_j) D_{m',m}^{\alpha*}(G_j) = \frac{g}{d_\alpha}\, \delta_{n,m}\delta_{n',m'} \qquad (A.36)$$

Finally, taking the complex conjugate of (A.36) and combining *Case 1* and *Case 2*, we obtain

$$\frac{d_\alpha}{g} \sum_{j=1}^{g} D_{n',n}^{\alpha*}(G_j) D_{m',m}^{\beta}(G_j) = \delta_{\alpha,\beta}\delta_{n',m'}\delta_{n,m} \qquad (A.37)$$

The great orthogonality relation is very central, and almost all of the chemically relevant results of representation theory depend upon it. For example, combining (A.16) with the definition of characters (A.14), we obtain:

$$\sum_{j=1}^{g} \chi^{\alpha*}(G_j)\chi^{\beta}(G_j) \equiv \sum_{j=1}^{g} \left\{ \sum_n D_{n,n}^{\alpha*}(G_j) \right\} \left\{ \sum_m D_{m,m}^{\beta}(G_j) \right\}$$
$$= \frac{g}{d_\alpha}\delta_{\alpha,\beta} \sum_n \sum_m \delta_{n,m}\delta_{n,m} = g\delta_{\alpha,\beta} \qquad (A.38)$$

Equation (A.37) holds only for unitary representations, but every representation is equivalent to a unitary representation since it is always possible to perform a similarity transformation that orthonormalizes the basis functions. Therefore, since characters are invariant under similarity transformations, the orthonormality of characters

$$\frac{1}{g} \sum_{j=1}^{g} \chi^{\alpha*}(G_j)\chi^{\beta}(G_j) = \delta_{\alpha,\beta} \equiv \begin{cases} 0 & \text{if the representations are inequivalent} \\ 1 & \text{if the representations are equivalent} \end{cases}$$

holds even for non-unitary irreducible representations.

Now consider a representation $D_{n',n}(G_j)$ which may be reducible. If we reduce it by means of a similarity transformation, then in its reduced form it will be block-diagonal, each block being irreducible. Taking the trace, we find that the character of an element in the reduced representation $D_{n',n}'(G_j)$ is the sum of the characters of the irreducible representations of which it is composed. Thus

$$\chi(G_j) \equiv \sum_{n=1}^{d} D_{n,n}'(G_j)$$
$$= \chi^1(G_j) + \chi^2(G_j) + \cdots \qquad (A.39)$$
$$= \sum_\beta n_\beta \chi^\beta(G_j)$$

where n_β is the number of times that the irreducible representation D^β occurs among the diagonal blocks of D'. Then from (A.39) we have

$$\frac{1}{g}\sum_{j=1}^{g}\chi^{\alpha*}(G_j)\chi(G_j) = \sum_{\beta}n_\beta\sum_{j=1}^{g}\chi^{\alpha*}(G_j)\chi^\beta(G_j)$$
$$= \sum_\beta n_\beta\delta_{\alpha,\beta} = n_\alpha \quad\quad\text{(A.40)}$$

This gives us a way to find out how many times a particular irreducible representation D^α occurs in a reducible representation D. According to (A.40), we just have to take the scalar product of the characters and divide by the order of the group. When we say that D^α 'occurs' n_α times in D, we mean that it is possible by means of a similarity transformation to bring D into block-diagonal form where D^α occurs n_α times along the diagonal blocks. The relationship is sometimes written in the form

$$D = n_1 D^1 + n_1 D^2 + \cdots \quad\quad\text{(A.41)}$$

Obviously in this decomposition we do not need to distinguish between different equivalent forms of an irreducible representation D^α, since all of them have the same character, and it is possible to go from one to another by means of a similarity transformation.

A.6 Classes

Two elements of a group G_i and G_j are said to be in the same 'class' if there exists another element G_l in the group such that

$$G_i = G_l^{-1}G_j G_l \quad\quad\text{(A.42)}$$

Thus, if we start with a particular element G_j, we can generate the set of elements in the same class by keeping j fixed in (A.42) and letting G_l run through all the elements of the group. It also follows from (A.42) that we can construct an operator M_k which commutes with all the elements of the group by summing the elements of a particular class:

$$M_k \equiv \sum_{\text{class } k} G_j \quad\quad\text{(A.43)}$$

Then for an arbitrary group element G_l we have

$$
\begin{aligned}
G_l^{-1}[M_k, G_l] &= \sum_{\text{class } k} G_l^{-1}[G_j, G_l] \\
&= \sum_{\text{class } k} \left(G_l^{-1}G_jG_l - G_j\right) \\
&= \sum_{\text{class } k} (G_i - G_j) = 0
\end{aligned}
\tag{A.44}
$$

Equation (A.44) can hold only if $[M_k, G_l] = 0$. An operator, such as M_k, which commutes with every element of the group is called an 'invariant'. If there are r classes in a group, there will be r linearly independent invariants that can be constructed in this way.

For any representation of two elements G_i and G_j in the same class, it follows from (A.42) that

$$
D(G_i) = D(G_l^{-1})D(G_j)D(G_l) = D(G_l)^{-1}D(G_j)D(G_l)
\tag{A.45}
$$

Thus if $D(G_i)$ and $D(G_i)$ represent two elements in the same class, they are connected by a similarity transformation, and therefore they have the same character. In other words, *all elements in the same class have the same character.* This means that in applying equation (A.40) we do not need to go through quite so much work. Instead of summing over all of the elements in the group, we can take the product of characters for a representative element in each class, multiply by the number of elements in the class, and then sum over the classes. If g_k represents the number of elements in the class k and r denotes the number of classes, then the orthogonality relation for characters, equation (A.39), can be written in the form

$$
\sum_{k=1}^{r} \sqrt{\frac{g_k}{g}} \chi_k^{\alpha*}(G_j) \sqrt{\frac{g_k}{g}} \chi_k^{\beta}(G_j) = \delta_{\alpha,\beta}
\tag{A.46}
$$

where χ_k^{α} is the character of a representative element in class k.

A.7   Projection operators

The great orthogonality theorem, equation (A.16), can be used to construct group-theoretical projection operators. Suppose that each of the sets of functions $(\Phi_1^1, \Phi_2^1, \ldots, \Phi_{d_1}^1)$, $(\Phi_1^2, \Phi_2^2, \ldots, \Phi_{d_2}^2)$, etc. forms the basis for an irreducible representation of a group, and that there are r' nonequivalent irreducible representations. Then

$$G_j \Phi_n^\beta = \sum_{n'=1}^{d_\beta} \Phi_{n'}^\beta D_{n',n}^\beta(G_j) \qquad (A.47)$$

Then from (A.16) we have

$$\sum_{j=1}^{g} D_{m,m}^{\alpha*}(G_j) G_j \Phi_n^\beta = \sum_{n'=1}^{d_\beta} \Phi_{n'}^\beta \sum_{j=1}^{g} D_{m,m}^{\alpha*}(G_j) D_{n',n}^\beta(G_j)$$

$$= \delta_{\alpha,\beta} \frac{g}{d_\alpha} \sum_{n'=1}^{d_\beta} \Phi_{n'}^\beta \delta_{m,n'} \delta_{m,n} \qquad (A.48)$$

$$= \delta_{\alpha,\beta} \frac{g}{d_\alpha} \Phi_m^\beta \delta_{m,n}$$

From (A.48) it follows that if we let

$$P_m^\alpha \equiv \frac{d_\alpha}{g} \sum_{j=1}^{g} D_{m,m}^{\alpha*}(G_j) G_j \qquad (A.49)$$

then

$$P_m^\alpha \Phi_n^\beta = \delta_{\alpha,\beta} \delta_{m,n} \Phi_m^\beta \qquad (A.50)$$

In other words, when the operator P_m^α defined by equation (A.49) acts on any function in the set $(\Phi_1^1, \Phi_2^1, \ldots, \Phi_{d_1}^1)$, $(\Phi_1^2, \Phi_2^2, \ldots, \Phi_{d_2}^2), \ldots$, the function is given back unchanged, provided that $m = n$ and $\alpha = \beta$. Otherwise the function is annihilated. *Thus, P_m^α is a projection operator corresponding to the mth basis function of the αth irreducible representation of the group in a standard unitary representation.* If P_m^α acts on an arbitrary function, it will annihilate all of it except the component that transforms like the mth basis function of D^α.

A second type of group-theoretical projection operator can be defined by the relationship

$$P^\alpha \equiv \sum_{m=1}^{d_\alpha} P_m^\alpha = \frac{d_\alpha}{g} \sum_{j=1}^{g} \sum_{m=1}^{d_\alpha} D_{m,m}^{\alpha*}(G_j) G_j \qquad (A.51)$$

which can be rewritten as

$$P^\alpha \equiv \frac{d_\alpha}{g} \sum_{j=1}^{g} \chi^{\alpha*}(G_j) G_j \qquad (A.52)$$

From (A.50) it follows that

$$P^\alpha \Phi_n^\beta = \sum_{m=1}^{d_\alpha} P_m^\alpha \Phi_n^\beta = \delta_{\alpha,\beta} \sum_{m=1}^{d_\alpha} \delta_{m,n} \Phi_m^\beta = \delta_{\alpha,\beta} \Phi_n^\beta \qquad (A.53)$$

When P^α acts on an arbitrary function, it annihilates everything except the component which can be expressed as a linear combination of basis functions of the irreducible representation D^α. If we sum (A.53) over all of the irreducible representations of the group, we obtain

$$\sum_{\alpha=1}^{r'} P^\alpha \Phi_n^\beta = \sum_{\alpha=1}^{r'} \delta_{\alpha,\beta} \Phi_n^\beta = \Phi_n^\beta \qquad (A.54)$$

Therefore the sum acts like the identity operator and we can write

$$\sum_{\alpha=1}^{r'} P^\alpha = E \qquad (A.55)$$

Combining (A.55) with (A.52), we obtain

$$\sum_{j=1}^{g} \sum_{\alpha=1}^{r'} \frac{d_\alpha}{g} \chi^{\alpha*}(G_j) G_j = E \equiv G_1 \qquad (A.56)$$

Since the group elements $G_1, \ldots, G_g$ are linearly independent, equation (A.55) implies that

$$\sum_{\alpha=1}^{r'} \frac{d_\alpha}{g} \chi^{\alpha*}(G_j) = \delta_{j,1} \qquad (A.57)$$

The character of the identity element in any representation is equal to the dimension of that representation:

$$\chi^{\alpha*}(E) = \chi^\alpha(E) = d_\alpha \qquad (A.58)$$

Therefore, when $j = 1$, we obtain from (A.57) the relationship

$$\sum_{\alpha=1}^{r'} d_\alpha^2 = g, \qquad (A.59)$$

i.e. *the sum of the squares of the dimensions of the irreducible representations is equal to the order of the group.*

A.8 The regular representation

The 'regular representation' of a finite group is a reducible representation D^{reg} in which the basis consists of the group elements themselves:

$$G_j G_n = \sum_{n'=1}^{g} G_{n'} D_{n',n}^{\mathrm{reg}}(G_j) \qquad (A.60)$$

D^{reg} must thus be a set of g $g \times g$ matrices. If we know the multiplication table for a finite group, we can construct the regular representation. For example, the multiplication table for the group C_3 is shown above. It can easily be verified that if we let

$$D^{\text{reg}}(E) = \begin{pmatrix} 1\,0\,0 \\ 0\,1\,0 \\ 0\,0\,1 \end{pmatrix}$$

$$D^{\text{reg}}(C_3) = \begin{pmatrix} 0\,0\,1 \\ 1\,0\,0 \\ 0\,1\,0 \end{pmatrix} \qquad \text{(A.61)}$$

$$D^{\text{reg}}(C_3^{-1}) = \begin{pmatrix} 0\,1\,0 \\ 0\,0\,1 \\ 1\,0\,0 \end{pmatrix}$$

then the matrices will be the regular representation of the group C_3 according to the definition shown in (A.60) and the multiplication table (A.1). Since $G_i G_j \neq G_j$ for $G_i \neq E$, it follows that the character of every group element except the identity element vanishes in the regular representation. (We can notice that this holds in the example given above.) Therefore in the case of the regular representation, equation (A.40) becomes:

$$n_\alpha = \frac{1}{g} \sum_{j=1}^{g} \chi^{\alpha*}(G_j)\chi^{\text{reg}}(G_j) = \frac{1}{g}\chi^{\alpha*}(E)\chi^{\text{reg}}(E) = d_\alpha \qquad \text{(A.62)}$$

Thus each irreducible representation of a finite group appears d_α times in the regular representation.

When each element of a group commutes with every other one, a group is said to be Abelian. Then from the definition of classes, (A.42), it follows that in an Abelian group, every element is in a class by itself, so that an Abelian group contains g classes, i.e. $r' = g$. We can next ask how many non-equivalent irreducible representations an Abelian group contains. To answer this question, we remember from Schur's lemma that the only matrix that commutes with every matrix in an irreducible representation of a group must be a multiple of the unit matrix. But in an Abelian group, all of the elements commute with each other, and therefore their irreducible representations must all be multiples of the unit matrix. This can happen only if all the irreducible representations are 1-dimensional. Thus for an Abelian group, $d_\alpha = 1$, $\alpha = 1, 2, \ldots, r'$ and $r' = g$. It can be seen from the multiplication table of the group C_3 that it is Abelian. In the example of C_3, (A.59) becomes $1 + 1 + 1 = 3$.

A.9   Classification of basis functions

We can use the group-theoretical projection operators to classify basis sets into basis functions for the various irreducible representations of a group. For example, we can construct the projection operators of the group C_3 from the character table:

$$P^1 = \frac{1}{3}\left(E + C_3 + C_3^{-1}\right)$$

$$P^2 = \frac{1}{3}\left(E + e^{-i2\pi/3}C_3 + e^{i2\pi/3}C_3^{-1}\right) \tag{A.63}$$

$$P^3 = \frac{1}{3}\left(E + e^{i2\pi/3}C_3 + e^{-i2\pi/3}C_3^{-1}\right)$$

Since the group C_3 is Abelian, all of its irreducible representations are 1-dimensional, and hence there is no difference between projection operators of the type P^α and those of the type P_n^α. Notice that $P^1 + P^2 + P^3 = E$ in accordance with (A.55), and that $P^\alpha P^\beta = \delta_{\alpha,\beta} P^\alpha$, as expected.

Now consider the set of functions $\Phi_m = e^{im\varphi}$ where m is an integer. We can use the projection operators of (A.63) to split the Hilbert space spanned by this set of functions into three subspaces. Using the relationships

$$Ee^{im\varphi} = e^{im\varphi}$$

$$C_3 e^{im\varphi} = e^{im(\varphi-2\pi/3)} \tag{A.64}$$

$$C_3^{-1} e^{im\varphi} = e^{im(\varphi+2\pi/3)}$$

we obtain

$$P^1 e^{im\varphi} = \frac{1}{3}e^{im\varphi}\left(1 + e^{-im2\pi/3} + e^{im2\pi/3}\right)$$

$$= \begin{cases} 0 & \text{if } m = \pm 1, \pm 2, \pm 4, \pm 5, \ldots \\ e^{im\varphi} & \text{if } m = 0, \pm 3, \pm 6, \pm 9, \ldots \end{cases} \tag{A.65}$$

and similarly

$$P^2 e^{im\varphi} = \begin{cases} 0 & \text{if } m + 1 = \pm 1, \pm 2, \pm 4, \pm 5, \ldots \\ e^{im\varphi} & \text{if } m + 1 = 0, \pm 3, \pm 6, \pm 9, \ldots \end{cases}$$

$$P^3 e^{im\varphi} = \begin{cases} 0 & \text{if } m - 1 = \pm 1, \pm 2, \pm 4, \pm 5, \ldots \\ e^{im\varphi} & \text{if } m - 1 = 0, \pm 3, \pm 6, \pm 9, \ldots \end{cases} \tag{A.66}$$

Thus the Hilbert space spanned by the functions $\Phi_m = e^{im\varphi}$ is divided into three subspaces each of which consists of basis functions for one of the irreducible representations of C_3. For non-Abelian groups the Hilbert space spanned by a set of basis functions can be divided into still smaller subspaces through the use of projection operators of the type P_n^α defined in equation (A.49). If we wish to have names for the two types of projection operators, we might call P_m^α 'strong' and P^α 'weak', since P_n^α has a stronger effect than P^α.

Now suppose that we have divided the Hilbert space spanned by a set of basis functions into small subspaces by means of the strong projection operators P_n^α, so that

$$P_n^\alpha \Phi_j = p_j \Phi_j \qquad p_j = 0 \text{ or } 1 \qquad (A.67)$$

We will now show that if an operator T commutes with every element of the group, then the matrix elements of T linking functions belonging to different subspaces must necessarily vanish. The proof is as follows: Since T commutes with every element of the group, and since the projection operators are constructed from group elements, we have

$$[P_n^\alpha, T] = 0 \qquad (A.68)$$

Then

$$\langle \Phi_j | [P_n^\alpha, T] | \Phi_k \rangle = (p_j - p_k) \langle \Phi_j | T | \Phi_k \rangle = 0 \qquad (A.69)$$

Thus if Φ_j and Φ_k belong to different subspaces when the basis set is classified by the action of the projection operators P_n^α, i.e. if $p_j \neq p_k$, then $\langle \Phi_j | T | \Phi_k \rangle = 0$. *It follows that a matrix representation of the operator T will be block-diagonal if it is based on functions that have been classified by means of the projection operators P_n^α, i.e. if it is based on a set of functions that satisfy (A.67). Such a basis set is said to be 'symmetry-adapted'.*

We can introduce a special notation to represent fully symmetry-adapted basis functions. Let $|\eta_j^{\alpha,n}\rangle$ be such a function.[1] By this we indicate that the function transforms under the action of the group elements like n^{th} basis function of the αth standard irreducible representation of the group, while the index j distinguishes between the various linearly independent functions that have this property. With this notation we can write:

$$P_n^\alpha |\eta_j^{\beta,m}\rangle = \delta_{\alpha,\beta} \delta_{n,m} |\eta_j^{\beta,m}\rangle \qquad (A.70)$$

Using this notation, the statement that a matrix representation of the operator T based on symmetry-adapted functions will be block-diagonal can be written in the form:

$$\left\langle \eta_i^{\alpha,n} \left| T \right| \eta_j^{\beta,m} \right\rangle = \delta_{\alpha,\beta} \delta_{n,m} \left\langle \eta_i^{\alpha,n} \left| T \right| \eta_j^{\beta,m} \right\rangle \qquad (A.71)$$

---

[1] We also introduce the Dirac notation here, since it is useful in the discussion of matrix elements.

The eigenvalues and eigenfunctions of T can also be expressed in this notation:

$$T\,|\Psi_\kappa^{\alpha,m}\rangle = \lambda_\kappa^{\alpha,m}\,|\Psi_\kappa^{\alpha,m}\rangle \qquad (A.72)$$

where

$$|\Psi_\kappa^{\alpha,m}\rangle = \sum_j |\eta_j^{\alpha,m}\rangle\, C_{j,\kappa} \qquad (A.73)$$

In other words, *a set of functions all of which transform like the nth basis function of the αth irreducible representation of a group combine to form an eigenfunction of an operator T that commutes with all of the group elements.*

We will now try to find a relationship between the degeneracy of the root $\lambda_\kappa^{\alpha,n}$ and the dimension d_α of the irreducible representation D^α. To do this, we introduce the 'shift operator'

$$P_{m',m}^\alpha \equiv \frac{d_\alpha}{g} \sum_{j=1}^{g} D_{m',m}^{\alpha*}(G_j) G_j \qquad m' \neq m \qquad (A.74)$$

Then by an argument similar to (A.48) we have

$$\begin{aligned}
P_{m',m}^\alpha |\eta_j^{\alpha,m}\rangle &= \frac{d_\alpha}{g} \sum_{j=1}^{g} D_{m',m}^{\alpha*}(G_j) G_j |\eta_j^{\alpha,m}\rangle \\
&= \sum_{m''=1}^{d_\alpha} |\eta_j^{\alpha,m''}\rangle \frac{d_\alpha}{g} \sum_{j=1}^{g} D_{m',m}^{\alpha*}(G_j) D_{m'',m}^{\alpha}(G_j) \\
&= \sum_{m''=1}^{d_\alpha} |\eta_j^{\alpha,m''}\rangle \delta_{m'',m'} = |\eta_j^{\alpha,m'}\rangle \qquad (A.75)
\end{aligned}$$

where we have made use of the great orthogonality relation (A.16). Since $P_{m',m}^\alpha$ is a linear combination of group elements, it must commute with T:

$$\left[P_{m',m}^\alpha, T\right] = 0 \qquad (A.76)$$

Therefore

$$\begin{aligned}
\left\langle \Psi_\kappa^{\alpha,m'} \left| \left[P_{m',m}^\alpha, T\right] \right| \Psi_\kappa^{\alpha,m} \right\rangle &= \left(\lambda_\kappa^{\alpha,m'} - \lambda_\kappa^{\alpha,m}\right) \left\langle \Psi_\kappa^{\alpha,m'} \left| P_{m',m}^\alpha \right| \Psi_\kappa^{\alpha,m} \right\rangle \\
&= \left(\lambda_\kappa^{\alpha,m'} - \lambda_\kappa^{\alpha,m}\right) = 0 \qquad (A.77)
\end{aligned}$$

so that the roots corresponding to the d_α eigenfunctions $|\Psi_\kappa^{\alpha,1}\rangle, \ldots, |\Psi_\kappa^{\alpha,d_\alpha}\rangle$ must be degenerate. Such a degeneracy is called a 'due degeneracy' because it is due to the symmetry properties of the system. If there are other degeneracies, they are termed 'accidental'.

Appendix B

STURMIAN BASIS SETS

B.1 One-electron Coulomb Sturmians

Because of their completeness properties, one-electron Sturmian basis sets have long been used in theoretical atomic physics [Shull and Löwdin, 1959], [Rotenberg, 1962], [Rotenberg, 1970], [Avery, 2003], [Gasaneo et al., 2009]. Their form is identical with that of the familiar hydrogenlike atomic orbitals, except that the factor Z/n is replaced by a constant k. The one-electron Coulomb Sturmians can be written as

$$\chi_{nlm}(\mathbf{x}) = R_{nl}(r)Y_{lm}(\theta, \phi) \tag{B.1}$$

where Y_{lm} is a spherical harmonic, and where the radial function has the form

$$R_{nl}(r) = \mathcal{N}_{nl}(2kr)^l e^{-kr} F\left(l + 1 - n | 2l + 2 | 2kr\right) \tag{B.2}$$

Here

$$\mathcal{N}_{nl} = \frac{2k^{3/2}}{(2l+1)!}\sqrt{\frac{(l+n)!}{n(n-l-1)!}} \tag{B.3}$$

is a normalizing constant, while

$$F\left(a|b|x\right) \equiv \sum_{t=0}^{\infty} \frac{a^{\bar{t}}}{t!b^{\bar{t}}}x^t = 1 + \frac{a}{b}x + \frac{a(a+1)}{2b(b+1)}x^2 + \cdots \tag{B.4}$$

is a confluent hypergeometric function. Coulomb Sturmian basis functions obey the following one-electron Schrödinger equation (in atomic units):

$$\left[-\frac{1}{2}\nabla^2 - \frac{nk}{r} + \frac{1}{2}k^2\right]\chi_{nlm}(\mathbf{x}) = 0 \tag{B.5}$$

which is just the Schrödinger equation for an electron in a hydrogenlike atom with the replacement $Z/n \to k$. All of the functions in such a basis set correspond to the same energy,

Table B.1: One-electron Coulomb Sturmian radial functions. If k is replaced by Z/n they are identical to the familiar hydrogenlike radial wave functions.

n	l	$R_{n,l}(r)$
1	0	$2k^{3/2}e^{-kr}$
2	0	$2k^{3/2}(1 - kr)e^{-kr}$
2	1	$\dfrac{2k^{3/2}}{\sqrt{3}}\, kr\, e^{-kr}$
3	0	$2k^{3/2}\left(1 - 2kr + \dfrac{2(kr)^2}{3}\right)e^{-kr}$
3	1	$2k^{3/2}\dfrac{2\sqrt{2}}{3}\, kr\left(1 - \dfrac{kr}{2}\right)e^{-kr}$
3	2	$2k^{3/2}\dfrac{\sqrt{2}}{3\sqrt{5}}\,(kr)^2\, e^{-kr}$

$$\epsilon = -\frac{1}{2}k^2 \qquad\qquad (\text{B.6})$$

In other words the basis set is isoenergetic. In the wave equation obeyed by the Sturmians, (B.5), the potential is weighted differently for members of the basis set corresponding to different values of n. Equation (B.5) can be written in the form:

$$\left[-\frac{1}{2}\nabla^2 - \beta_n\frac{Z}{r} + \frac{1}{2}k^2\right]\chi_{nlm}(\mathbf{x}) = 0 \qquad \beta_n = \frac{kn}{Z} \qquad (\text{B.7})$$

The weighting factors β_n are chosen in such a way as to make all of the solutions isoenergetic. All solutions correspond to the energy $\epsilon = -k^2/2$.

In the Hamiltonian formulation of physics, the eigenvalues of the wave equation are a spectrum of allowed energies, but here all of the solutions of the wave equation correspond to the same energy, and the weighting factors play the role of eigenvalues. The functions in a Coulomb Sturmian basis set can be shown to obey a potential-weighted orthonormality relation: To see this, we consider two solutions, $\chi_{nlm}(\mathbf{x})$ and $\chi_{n'l'm'}(\mathbf{x})$, obeying the equations:

$$\left[-\frac{1}{2}\nabla^2 + \frac{1}{2}k^2\right]\chi_{nlm}(\mathbf{x}) = \frac{nk}{r}\chi_{nlm}(\mathbf{x})$$

$$\left[-\frac{1}{2}\nabla^2 + \frac{1}{2}k^2\right]\chi_{n'l'm'}^*(\mathbf{x}) = \frac{n'k}{r}\chi_{n'l'm'}^*(\mathbf{x})$$

(B.8)

Multiplying the two equations from the left respectively by $\chi_{n'l'm'}^*(\mathbf{x})$ and $\chi_{nlm}(\mathbf{x})$, integrating over the coordinates, and subtracting the two equations, we obtain:

$$(n - n')\int d^3x \ \chi_{n'l'm'}^*(\mathbf{x})\frac{1}{r}\chi_{nlm}(\mathbf{x}) = 0 \tag{B.9}$$

where we have also made use of the fact that (from Hermiticity)

$$\int d^3x \ \chi_{n'l'm'}^*(\mathbf{x})\left[-\frac{1}{2}\nabla^2 + \frac{1}{2}k^2\right]\chi_{nlm}(\mathbf{x})$$

$$- \int d^3x \ \chi_{nlm}(\mathbf{x})\left[-\frac{1}{2}\nabla^2 + \frac{1}{2}k^2\right]\chi_{n'l'm'}^*(\mathbf{x}) = 0$$

(B.10)

Thus for $n \neq n'$, the potential-weighted scalar product vanishes, and it vanishes also when $l' \neq l$ or $m' \neq m$ because of the orthogonality of the spherical harmonics. The Coulomb Sturmians are normalized in such a way that the orthonormality relation is:

$$\int d^3x \ \chi_{n'l'm'}^*(\mathbf{x})\frac{1}{r}\chi_{nlm}(\mathbf{x}) = \frac{k}{n}\delta_{n'n}\delta_{l'l}\delta_{m'm} \tag{B.11}$$

Because of their completeness and their close relationship with Coulomb potentials, Coulomb Sturmians are widely used in atomic physics.

B.2 Löwdin-orthogonalized Coulomb Sturmians

The Coulomb Sturmians form a complete set in the sense that any square-integrable function of $\mathbf{x}$ can be expanded in terms of them. For this reason, they are useful as basis functions in many applications. Sometimes it may be convenient to use Coulomb Sturmian basis functions in a form that is orthonormalized in the conventional way. Let us denote the orthogonalized Coulomb Sturmians by $\tilde{\chi}_\mu(\mathbf{x})$, where $\mu \equiv (n, l, m)$. This new basis set is related to the original set of Coulomb Sturmians discussed above by

$$\tilde{\chi}_\mu(\mathbf{x}) = \sum_{\mu'} \chi_\mu(\mathbf{x}) W_{\mu',\mu} \tag{B.12}$$

where $W_{\mu',\mu}$ is a transformation matrix. We wish the transformation to be such that

$$\int d^3x \; \tilde{\chi}_{\mu'}^*(\mathbf{x})\tilde{\chi}_\mu(\mathbf{x}) \equiv \tilde{S}_{\mu',\mu} = \delta_{\mu',\mu} \tag{B.13}$$

Suppose that

$$\int d^3x \; \chi_{\mu'}^*(\mathbf{x})\chi_\mu(\mathbf{x}) = S_{\mu',\mu} \tag{B.14}$$

Then, in matrix notation, the condition that the transformation matrix W must satisfy is

$$W^\dagger S W = I \tag{B.15}$$

where the dagger denotes the Hermitian adjoint, i.e. the conjugate transpose. Following Löwdin and Wannier, we can choose from all the possible solutions to the matrix equation (B.15) the one for which

$$W^\dagger = W \tag{B.16}$$

(This is sometimes called *symmetrical orthogonalization.*) Then (B.15) will be satisfied if

$$W = S^{-1/2} \tag{B.17}$$

In order to find the square root of the overlap matrix S, we diagonalize it, take the inverse square root in the diagonal representation, and then transform back to the original representation. This gives us $W = S^{-1/2}$, which we then use to perform the transformation shown in equation (B.12).

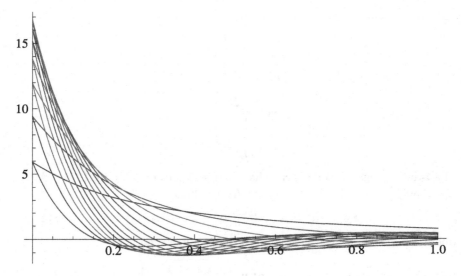

Fig. B.1 *A set of 15 Löwdin-orthogonalized Coulomb Sturmians corresponding to $l = 0$ and $k = 1$. The radial parts are shown as functions of r. If an arbitrary radial function is to be expanded in terms of this set, the value of k for the set can be adjusted in such a way as to give maximum accuracy. Löwdin-orthogonalized Coulomb Sturmians are used in the Hartree-Fock calculations of Chapter 2.*

B.3　The Fock projection

Coulomb Sturmian basis functions and their Fourier transforms are related by

$$\chi_{nlm}(\mathbf{x}) = \frac{1}{\sqrt{(2\pi)^3}} \int d^3x \; e^{i\mathbf{p}\cdot\mathbf{x}} \chi_{nlm}^t(\mathbf{p}) \qquad (B.18)$$

and by the inverse transform

$$\chi_{nlm}^t(\mathbf{p}) = \frac{1}{\sqrt{(2\pi)^3}} \int d^3x \; e^{-i\mathbf{p}\cdot\mathbf{x}} \chi_{nlm}(\mathbf{x}) \qquad (B.19)$$

By projecting momentum-space onto the surface of a 4-dimensional hypersphere, V. Fock [Fock, 1935], [Fock, 1958] was able to show that the Fourier-transformed Coulomb Sturmians can be very simply expressed in terms of 4-dimensional hyperspherical harmonics through the relationship

$$\chi_{n,l,m}^t(\mathbf{p}) = M(p) Y_{n-1,l,m}(\hat{\mathbf{u}}) \qquad (B.20)$$

where

$$M(p) \equiv \frac{4k^{5/2}}{(k^2 + p^2)^2} \qquad (B.21)$$

and

$$u_1 = \frac{2kp_1}{k^2 + p^2}$$

$$u_2 = \frac{2kp_2}{k^2 + p^2}$$

$$u_3 = \frac{2kp_3}{k^2 + p^2} \qquad \text{(B.22)}$$

$$u_4 = \frac{k^2 - p^2}{k^2 + p^2}$$

The 4-dimensional hyperspherical harmonics are given by [Judd, 1975], [Avery, 1989], [Avery, 2000], [Avery and Avery, 2006]

$$Y_{\lambda,l,m}(\hat{\mathbf{u}}) = \mathcal{N}_{\lambda,l} C_{\lambda-l}^{1+l}(u_4) Y_{l,m}(u_1, u_2, u_3) \qquad \text{(B.23)}$$

where $Y_{l,m}$ is a spherical harmonic of the familiar type, while

$$\mathcal{N}_{\lambda,l} = (-1)^{\lambda} i^l (2l)!! \sqrt{\frac{2(\lambda+1)(\lambda-l)!}{\pi(\lambda+l+1)!}} \qquad \text{(B.24)}$$

is a normalizing constant, and

$$C_j^{\alpha}(u_4) = \sum_{t=0}^{[j/2]} \frac{(-1)^t \Gamma(j+\alpha-t)}{t!(j-2t)!\Gamma(\alpha)} (2u_4)^{j-2t} \qquad \text{(B.25)}$$

is a Gegenbauer polynomial. The relationships between hyperspherical harmonics, harmonic polynomials, and harmonic projection will be discussed in Appendix C. Table 5.1 in Chapter 5 shows the first few hyperspherical harmonics.

B.4   Generalized Sturmians and many-particle problems

In 1968, Osvaldo Goscinski [Goscinski, 1968, 2003] generalized the concept of Sturmian basis sets by considering isoenergetic sets of solutions to a many-particle Schrödinger equation with a weighted potential:

$$\left[-\frac{1}{2}\Delta + \beta_{\nu} V_0(\mathbf{x}) - E_{\kappa} \right] |\Phi_{\nu}\rangle = 0 \qquad \text{(B.26)}$$

The weighting factors β_{ν} are chosen in such a way as to make all of the functions in the set correspond to the same energy, E_{κ}, and this energy is usually chosen to be that of the quantum mechanical state which is to be represented by a superposition of generalized Sturmian basis functions. If the basis set is used to treat N-particle systems where the particles have different masses, the operator Δ in the kinetic energy term is given by

$$\Delta \equiv \sum_{j=1}^{N} \frac{1}{m_j} \nabla_j^2 \qquad \text{(B.27)}$$

while if the masses are all equal, it is given by the generalized Laplacian operator:

$$\Delta \equiv \sum_{j=1}^{d} \frac{\partial^2}{\partial x_j^2} \qquad \text{(B.28)}$$

with $d = 3N$ and

$$\mathbf{x} = (x_1, x_2, \ldots, x_d) \qquad \text{(B.29)}$$

Like the one-electron Coulomb Sturmians, the functions in generalized Sturmian basis sets can be shown to obey a potential-weighted orthonormality relation [Avery and Avery, 2006]:

$$\langle \Phi_{\nu'} | V_0(\mathbf{x}) | \Phi_\nu \rangle = \delta_{\nu',\nu} \frac{2E_\kappa}{\beta_\nu} \qquad \text{(B.30)}$$

B.5 Use of generalized Sturmian basis sets to solve the many-particle Schrödinger equation

If we wish to solve a many-particle Schrödinger equation of the form

$$\left[-\frac{1}{2}\Delta + V(\mathbf{x}) - E_\kappa \right] |\Psi_\kappa\rangle = 0 \qquad \text{(B.31)}$$

we can approximate a solution as a superposition of generalized Sturmian basis functions

$$|\Psi_\kappa\rangle \approx \sum_\nu |\Phi_\nu\rangle B_{\nu,\kappa} \qquad \text{(B.32)}$$

Substituting this superposition into the Schrödinger equation and remembering that each of the basis functions satisfies (B.26), we obtain:

$$\sum_\nu \left[-\frac{1}{2}\Delta + V(\mathbf{x}) - E_\kappa \right] |\Phi_\nu\rangle B_{\nu,\kappa}$$
$$= \sum_\nu [V(\mathbf{x}) - \beta_\nu V_0(\mathbf{x})] |\Phi_\nu\rangle B_{\nu,\kappa} \approx 0 \qquad \text{(B.33)}$$

If we multiply from the left by a conjugate function from our generalized Sturmian basis set and integrate over all coordinates, we obtain a set of secular equations from which the kinetic energy term has disappeared:

$$\sum_{\nu} \langle \Phi_{\nu'}^* | \left[V(\mathbf{x}) - \beta_\nu V_0(\mathbf{x}) \right] \Phi_\nu \rangle B_{\nu,\kappa} = 0 \qquad \text{(B.34)}$$

If we introduce the definition

$$T_{\nu',\nu} \equiv -\frac{1}{p_\kappa} \langle \Phi_{\nu'}^* | V(\mathbf{x}) | \Phi_\nu \rangle \qquad \text{(B.35)}$$

where

$$p_\kappa \equiv \sqrt{-2E_\kappa} \qquad \text{(B.36)}$$

and make use of the potential-weighted orthonormality relations (B.30), we can rewrite the secular equations in the form:

$$\sum_{\nu} \left[T_{\nu',\nu} - p_\kappa \delta_{\nu',\nu} \right] B_{\nu,\kappa} = 0 \qquad \text{(B.37)}$$

The generalized Sturmian secular equations are strikingly different from conventional Hamiltonian secular equations in several ways:

- The kinetic energy term has disappeared.
- The matrix representing the approximate potential $V_0(\mathbf{x})$ is diagonal.
- The roots of the secular equations are not energies, but values of the scaling parameter p_κ, from which the energy can be obtained through the relationship $E_\kappa = -p_\kappa^2/2$.
- For Coulomb potentials, the matrix $T_{\nu',\nu}$ is energy-independent.

B.6   Momentum-space orthonormality relations for Sturmian basis sets

By arguments similar to those used in equations (B.8)-(B.11), a set of generalized Sturmian basis functions can be shown to obey a potential-weighted orthonormality relation in direct space

$$\int d\mathbf{x} \ \Phi_{\nu'}^*(\mathbf{x}) V_0(\mathbf{x}) \Phi_\nu(\mathbf{x}) = \delta_{\nu',\nu} \frac{2E_\kappa}{\beta_\nu} = -\delta_{\nu',\nu} \frac{p_\kappa^2}{\beta_\nu} \qquad \text{(B.38)}$$

where

$$p_\kappa^2 \equiv -2E_\kappa \qquad \text{(B.39)}$$

(in equation (B.38) and in the remainder of this appendix, we abandon the Dirac bra and ket notation in order to distinguish between functions of $\mathbf{x} \equiv (\mathbf{x}_1, \mathbf{x}_2, \ldots, \mathbf{x}_N)$ and functions of $\mathbf{p} \equiv (\mathbf{p}_1, \mathbf{p}_2, \ldots, \mathbf{p}_N))$. We would now like to find the momentum-space orthonormality relations obeyed by Fourier transforms of the generalized Sturmian basis set. Because the Fourier transform is unitary, the inner product of any two functions in L_2 is preserved under the operation of taking their Fourier transforms, i.e.

$$\int dx \; f^*(\mathbf{x})g(\mathbf{x}) = \int dp \; f^{t*}(\mathbf{p})g^t(\mathbf{p}) \tag{B.40}$$

Using this well-known relationship with $f^*(\mathbf{x}) = \Phi_{\nu'}^*(\mathbf{x})$ and $g(\mathbf{x}) = V_0(\mathbf{x})\Phi_\nu(\mathbf{x})$, we have

$$\int dx \; \Phi_{\nu'}^*(\mathbf{x})V_0(\mathbf{x})\Phi_\nu(\mathbf{x}) = \int dp \; \Phi_{\nu'}^{t*}(\mathbf{p}) \left[V_0\Phi_\nu\right]^t (\mathbf{p}) \tag{B.41}$$

In order to evaluate $[V_0\Phi_\nu]^t (\mathbf{p})$, we remember the Fourier convolution theorem, which states that the Fourier transform of the product of two functions is the convolution of their Fourier transforms. Thus if a and b are any two functions in L_2,

$$[ab]^t (\mathbf{p}') \equiv \frac{1}{(2\pi)^{d/2}} \int dx \; e^{-i\mathbf{p}'\cdot\mathbf{x}}a(\mathbf{x})b(\mathbf{x}) = \frac{1}{(2\pi)^{d/2}} \int dp \; a^t(\mathbf{p}'-\mathbf{p})b^t(\mathbf{p}) \tag{B.42}$$

Letting $a(\mathbf{x}) = V_0(\mathbf{x})$ and $b(\mathbf{x}) = \Phi_\nu(\mathbf{x})$ we have

$$[V_0\Phi_\nu]^t (\mathbf{p}') = \frac{1}{(2\pi)^{d/2}} \int dp \; V_0^t(\mathbf{p}'-\mathbf{p})\Phi_\nu^t(\mathbf{p}) \tag{B.43}$$

Since the momentum-space integral equation corresponding to (B.26) has the form

$$(p'^2 + p_\kappa^2)\Phi_\nu^t(\mathbf{p}') = -\frac{2\beta_\nu}{(2\pi)^{d/2}} \int dp \; V_0^t(\mathbf{p}'-\mathbf{p})\Phi_\nu^t(\mathbf{p}) \tag{B.44}$$

it follows that

$$[V_0\Phi_\nu]^t (\mathbf{p}) = -\frac{(p^2 + p_\kappa^2)}{2\beta_\nu}\Phi_\nu^t(\mathbf{p}) \tag{B.45}$$

Finally, substituting (B.45) into (B.41), we obtain the momentum-space orthonormality relations for a set of generalized Sturmian basis functions:

$$\int dp \; \Phi_{\nu'}^{t*}(\mathbf{p}) \left(\frac{p^2 + p_\kappa^2}{2p_\kappa^2}\right) \Phi_\nu^t(\mathbf{p}) = \delta_{\nu',\nu} \tag{B.46}$$

Because all of the functions $\Phi_\nu(\mathbf{x})$ in the generalized Sturmian basis set obey equation (B.26), the potential-weighted direct space orthonormality relations shown in equation (B.38) can be rewritten in the form

$$\int dx \; \Phi_{\nu'}^*(\mathbf{x}) \left(\frac{-\Delta + p_\kappa^2}{2p_\kappa^2}\right) \Phi_\nu(\mathbf{x}) = \delta_{\nu',\nu} \tag{B.47}$$

so that the momentum-space and direct-space orthonormality relations can be seen to be related to each other in a symmetrical way. These weighted orthonormality relations in $L_2(\mathbb{R}^d)$ are the usual orthonormality relations in the Sobolev space $W_2^{(1)}(\mathbb{R}^d)$ (see [Weniger, 1985]). For the case of unequal masses, where

$$\Delta \equiv \sum_{j=1}^{d} \frac{1}{m_j} \frac{\partial^2}{\partial x_j^2} \tag{B.48}$$

the momentum-space orthonormality relations for generalized Sturmians (B.46) takes on the slightly modified form

$$\int dp \; \Phi_{\nu'}^{t*}(\mathbf{p}) \left(\frac{\sum_j p_j^2/m_j + p_\kappa^2}{2p_\kappa^2} \right) \Phi_\nu^t(\mathbf{p}) = \delta_{\nu',\nu} \tag{B.49}$$

B.7 Sturmian expansions of d-dimensional plane waves

If the set of generalized Sturmian basis functions is complete in the sense of spanning the Sobolev space $W_2^{(1)}(\mathbb{R}^d)$, we can use it to construct a weakly convergent expansion of a d-dimensional plane wave (valid only in the sense of distributions). Suppose that we let

$$e^{i\mathbf{p}\cdot\mathbf{x}} = \left(\frac{p_\kappa^2 + p^2}{2p_\kappa^2} \right) \sum_\nu \Phi_\nu^{t*}(\mathbf{p}) a_\nu(\mathbf{x}) \tag{B.50}$$

We can then determine the unknown functions $a_\nu(\mathbf{x})$ by means of the orthonormality relations (B.46). Multiplying (B.50) on the left by $\Phi_{\nu'}^{t*}(\mathbf{p})$ and integrating over dp making use of (B.46), we obtain

$$\int dp \; e^{i\mathbf{p}\cdot\mathbf{x}} \Phi_{\nu'}^{t*}(\mathbf{p}) = \sum_\nu \delta_{\nu',\nu} a_\nu(\mathbf{x}) = a_{\nu'}(\mathbf{x}) \tag{B.51}$$

so that

$$a_\nu(\mathbf{x}) = \int dp \; e^{i\mathbf{p}\cdot\mathbf{x}} \Phi_\nu^t(\mathbf{p}) = (2\pi)^{d/2} \Phi_\nu(\mathbf{x}) \tag{B.52}$$

Thus finally we obtain an expansion of the form

$$e^{i\mathbf{p}\cdot\mathbf{x}} = (2\pi)^{d/2} \left(\frac{p_\kappa^2 + p^2}{2p_\kappa^2} \right) \sum_\nu \Phi_\nu^{t*}(\mathbf{p}) \Phi_\nu(\mathbf{x}) \tag{B.53}$$

If the set of generalized Sturmians $\Phi_\nu(\mathbf{x})$ does not span $W_2^{(1)}(\mathbb{R}^d)$, equation (B.53) becomes

$$P\left[e^{i\mathbf{p}\cdot\mathbf{x}} \right] = (2\pi)^{d/2} \left(\frac{p_\kappa^2 + p^2}{2p_\kappa^2} \right) \sum_\nu \Phi_\nu^{t*}(\mathbf{p}) \Phi_\nu(\mathbf{x}) \tag{B.54}$$

where $P\left[e^{i\mathbf{p}\cdot\mathbf{x}}\right]$ is the projection of the d-dimensional plane wave onto the subspace spanned by the set $\{\Phi_\nu(\mathbf{x})\}$. For example, if we are considering a system of N electrons, with $d = 3N$, the generalized Sturmian basis set might be antisymmetric with respect to exchange of the N electron coordinates but otherwise complete. In that case, $P\left[e^{i\mathbf{p}\cdot\mathbf{x}}\right]$ would represent the projection of the plane wave onto that part of Hilbert space corresponding to functions of $\mathbf{x}$ that are antisymmetric with respect to exchange of the N electron coordinates. Neither the expansion shown in equation (B.53) nor that shown in equation (B.54) is point-wise convergent. In other words, we cannot perform the sums shown on the right-hand sides of these equations and expect them to give point-wise convergent representations of the plane wave or its projection. However, the expansions are valid in the sense of distributions. For the case of unequal masses, the generalized Sturmian plane wave expansion takes on the slightly modified form

$$e^{i\mathbf{p}\cdot\mathbf{x}} = (2\pi)^{d/2}\left(\frac{p_\kappa^2 + \sum_j p_j^2/m_j}{2p_\kappa^2}\right)\sum_\nu \Phi_\nu^{t*}(\mathbf{p})\Phi_\nu(\mathbf{x}) \tag{B.55}$$

B.8 An alternative expansion of a d-dimensional plane wave

In the Hamiltonian formulation of physics, one typically obtains sets of functions whose orthonormality relation has the form

$$\int dx\ \Phi_{\nu'}^*(\mathbf{x})\Phi_\nu(\mathbf{x}) = \delta_{\nu',\nu} \tag{B.56}$$

Such a set of basis functions might, for example be the configurations resulting from the solution of the N-electron approximate Schrödinger equation

$$\left[-\frac{1}{2}\Delta + V_0(\mathbf{x}) - E_\nu\right]\Phi_\nu(\mathbf{x}) = 0 \tag{B.57}$$

with $\mathbf{x} \equiv (x_1, x_2, \ldots, x_d)$ and $d = 3N$. It is interesting to notice that a d-dimensional plane wave can also be expanded in terms of a basis set with orthonormality relations of the form shown in equation (B.56). To see this we write

$$e^{-i\mathbf{p}\cdot\mathbf{x}} = \sum_\nu a_\nu(\mathbf{p})\Phi_\nu^*(\mathbf{x}) \tag{B.58}$$

Multiplying from the left by $\Phi_{\nu'}(\mathbf{x})$ and integrating over the coordinates, we obtain the relation

$$\int dx\ e^{-i\mathbf{p}\cdot\mathbf{x}}\Phi_{\nu'}(\mathbf{x}) = \sum_\nu a_\nu(\mathbf{p})\int dx\ \Phi_\nu^*(\mathbf{x})\Phi_{\nu'}(\mathbf{x})$$

$$= \sum_\nu a_\nu(\mathbf{p})\delta_{\nu',\nu} = a_{\nu'}(\mathbf{p}) = (2\pi)^{d/2}\Phi_{\nu'}^t(\mathbf{p}) \tag{B.59}$$

Thus we obtain the alternative expansion

$$e^{-i\mathbf{p}\cdot\mathbf{x}} = (2\pi)^{d/2} \sum_{\nu} \Phi_{\nu'}^{t}(\mathbf{p})\Phi_{\nu}^{*}(\mathbf{x}) \tag{B.60}$$

or

$$e^{i\mathbf{p}\cdot\mathbf{x}} = (2\pi)^{d/2} \sum_{\nu} \Phi_{\nu'}^{t*}(\mathbf{p})\Phi_{\nu}(\mathbf{x}) \tag{B.61}$$

The expansion (B.53) was a consequence of the weighted orthonormality relations obeyed by generalized Sturmian basis sets, while the expansion (B.61) resulted from the more conventional orthonormality relations (B.56). Both forms of the expansion are used in Chapter 8.

Appendix C

ANGULAR AND HYPERANGULAR INTEGRATION

The physical importance of the properties of homogeneous and harmonic polynomials comes from their close relationship to spherical and hyperspherical harmonics, and from their relationship to angular and hyperangular integrations. We will see that these properties lead to useful results in both atomic and molecular physics.

C.1 Monomials, homogeneous polynomials, and harmonic polynomials

A *monomial of degree n* in d coordinates is a product of the form

$$m_n = x_1^{n_1} x_2^{n_2} x_3^{n_3} \cdots x_d^{n_d} \tag{C.1}$$

where the n_j's are positive integers or zero and where their sum is equal to n.

$$n_1 + n_2 + \cdots + n_d = n \tag{C.2}$$

For example, x_1^3, $x_1^2 x_2$ and $x_1 x_2 x_3$ are all monomials of degree 3. Since

$$\frac{\partial m_n}{\partial x_j} = n_j x_j^{-1} m_n \tag{C.3}$$

it follows that

$$\sum_{j=1}^{d} x_j \frac{\partial m_n}{\partial x_j} = n m_n \tag{C.4}$$

A *homogeneous polynomial of degree n* (which we will denote by the symbol f_n) is a series consisting of one or more monomials, all of which have degree n. For example, $f_3 = x_1^3 + x_1^2 x_2 - x_1 x_2 x_3$ is a homogeneous polynomial

161

of degree 3. Since each of the monomials in such a series obeys (C.4), it follows that

$$\sum_{j=1}^{d} x_j \frac{\partial f_n}{\partial x_j} = n f_n \qquad (C.5)$$

This simple relationship has very far-reaching consequences: If we now introduce the generalized Laplacian operator

$$\Delta \equiv \sum_{j=1}^{d} \frac{\partial^2}{\partial x_j^2} \qquad (C.6)$$

and the hyperradius defined by

$$r^2 \equiv \sum_{j=1}^{d} x_j^2 \qquad (C.7)$$

we can show (with a certain amount of effort!) that

$$\Delta \left(r^\beta f_\alpha \right) = \beta(\beta + d + 2\alpha - 2) r^{\beta-2} f_\alpha + r^\beta \Delta f_\alpha \qquad (C.8)$$

where α and β are positive integers or zero, β being even. We next define a *harmonic polynomial of degree n* to be a homogeneous polynomial of degree n which also satisfies the generalized Laplace equation:

$$\Delta h_n = 0 \qquad (C.9)$$

For example, $h_3 = x_1^2 x_2 - x_3^2 x_2 + x_1 x_2 x_3$ is a harmonic polynomial of degree 3. Combining (C.8) and (C.9) we obtain

$$\Delta \left(r^\beta h_\alpha \right) = \beta(\beta + d + 2\alpha - 2) r^{\beta-2} h_\alpha \qquad (C.10)$$

C.2 The canonical decomposition of a homogeneous polynomial

Every homogeneous polynomial f_n can be decomposed into a sum of harmonic polynomials multiplied by powers of the hyperradius. This decomposition, which is called *the canonical decomposition of a homogeneous polynomial*, has the form [Vilenkin, 1968]:

$$f_n = h_n + r^2 h_{n-2} + r^4 h_{n-4} + \cdots \qquad (C.11)$$

To see how the decomposition may be performed, we can act on both sides of equation (C.11) with the generalized Laplacian operator Δ. If we do this several times, making use of (C.10), we obtain [Avery, 1989], [Avery, 2000], [Avery and Avery, 2006]:

$$\Delta f_n = 2(d + 2n - 4)h_{n-2} + 4(d + 2n - 6)r^2 h_{n-4} + \cdots$$

$$\Delta^2 f_n = 8(d + 2n - 6)(d + 2n - 8)h_{n-4} + \cdots \tag{C.12}$$

$$\Delta^3 f_n = 48(d - 2n - 8)(d - 2n - 10)(d - 2n - 12)h_{n-6} + \cdots$$

and in general

$$\Delta^\nu f_n = \sum_{k=\nu}^{\lfloor \frac{n}{2} \rfloor} \frac{(2k)!!}{(2k - 2\nu)!!} \frac{(d + 2n - 2k - 2)!!}{(d + 2n - 2k - 2\nu - 2)!!} r^{2k-2\nu} h_{n-2k} \tag{C.13}$$

where

$$j!! \equiv \begin{cases} j(j-2)(j-4)\cdots 4 \times 2 & j = \text{even} \\ \\ j(j-2)(j-4)\cdots 3 \times 1 & j = \text{odd} \end{cases} \tag{C.14}$$

$$0!! \equiv 1$$

$$(-1)!! \equiv 1$$

An important special case occurs when $\nu = n/2$. In that case, (C.13) becomes

$$\Delta^{n/2} f_n = \frac{n!!(d + n - 2)!!}{(d - 2)!!} h_0 \tag{C.15}$$

or

$$h_0 = \frac{(d - 2)!!}{n!!(d + n - 2)!!} \Delta^{n/2} f_n \tag{C.16}$$

We will see below that this result leads to powerful angular and hyperangular integration theorems.

C.3 Harmonic projection

Equations (C.12) or (C.13) form a set of simultaneous equations that can be solved to yield expressions for the various harmonic polynomials that occur in the canonical decomposition of a homogeneous polynomial f_n. In this way we obtain the general result:

$$h_{n-2\nu} = \frac{(d + 2n - 4\nu - 2)!!}{(2\nu)!!(d + 2n - 2\nu - 2)!!}$$

$$\times \sum_{j=0}^{\lfloor \frac{n}{2} - \nu \rfloor} \frac{(-1)^j (d + 2n - 4\nu - 2j - 4)!!}{(2j)!!(d + 2n - 4\nu - 4)!!} r^{2j} \Delta^{j+\nu} f_n \tag{C.17}$$

If we let $n - 2\nu = \lambda$, this becomes

$$O_\lambda[f_n] = h_\lambda = \frac{(d + 2\lambda - 2)!!}{(n - \lambda)!!(d + n + \lambda - 2)!!}$$

$$\times \sum_{j=0}^{\lfloor \lambda/2 \rfloor} \frac{(-1)^j (d + 2\lambda - 2j - 4)!!}{(2j)!!(d + 2\lambda - 4)!!} r^{2j} \Delta^{j + \frac{1}{2}(n-\lambda)} f_n \quad (C.18)$$

Here O_λ can be thought of as a projection operator that projects out the harmonic polynomial of degree λ from the canonical decomposition of the homogeneous polynomial f_n. The projection is of course taken to be zero if λ and n have different parities or if $\lambda > n$.

C.4   Generalized angular momentum

The generalized angular momentum operator Λ^2 is defined as

$$\Lambda^2 \equiv -\sum_{i>j}^{d} \left(x_i \frac{\partial}{\partial x_j} - x_j \frac{\partial}{\partial x_i} \right)^2 \quad (C.19)$$

When $d = 3$, it reduces to the familiar orbital angular momentum operator

$$L^2 = L_1^2 + L_2^2 + L_3^2 \quad (C.20)$$

where

$$L_1 = \frac{1}{i} \left(x_2 \frac{\partial}{\partial x_3} - x_3 \frac{\partial}{\partial x_2} \right) \quad (C.21)$$

and where L_2 and L_3 given by similar expressions with cyclic permutations of the coordinates. If we expand the expression in (C.19), we obtain:

$$\Lambda^2 = -r^2 \Delta + \sum_{i,j=1}^{d} x_i x_j \frac{\partial^2}{\partial x_i \partial x_j} + (d - 1) \sum_{i=1}^{d} x_i \frac{\partial}{\partial x_i} \quad (C.22)$$

We next allow Λ^2 to act on a homogeneous polynomial f_n, and make use of (C.5). This gives us:

$$\Lambda^2 f_n = -r^2 \Delta f_n + n(d - 1) f_n + \sum_{i,j=1}^{d} x_i x_j \frac{\partial^2 f_n}{\partial x_i \partial x_j} \quad (C.23)$$

The relationship

$$\sum_{i,j=1}^{d} x_i x_j \frac{\partial^2 f_n}{\partial x_i \partial x_j} = n(n - 1) f_n \quad (C.24)$$

can be derived in a manner similar to the derivation of (C.5). Substituting this into (C.24) we have:

$$\Lambda^2 f_n = -r^2 \Delta f_n + n(n + d - 2)f_n \tag{C.25}$$

From (C.25) it follows that a harmonic polynomial of degree n is an eigenfunction of the generalized angular momentum operator with the eigenvalue $n(n + d - 2)$, i.e.

$$\Lambda^2 h_n = n(n + d - 2)h_n \tag{C.26}$$

It is conventional to use the symbol λ for the degree of a harmonic polynomial. Written in this notation, we have

$$\Lambda^2 h_\lambda = \lambda(\lambda + d - 2)h_\lambda \tag{C.27}$$

When $d = 3$, this reduces to

$$L^2 h_l = l(l + 1)h_l \tag{C.28}$$

We can conclude from this discussion that the canonical decomposition of a homogeneous polynomial can be viewed as a decomposition into eigenfunctions of generalized angular momentum.

C.5   Angular and hyperangular integration

In a 3-dimensional space the volume element is given by $dx_1 dx_2 dx_3$ in Cartesian coordinates or by $r^2 dr \, d\Omega$ in spherical polar coordinates. Thus we can write

$$dx_1 dx_2 dx_3 = r^2 dr \, d\Omega \tag{C.29}$$

where $d\Omega$ is the element of solid angle. Similarly, in a d-dimensional space we can write

$$dx_1 dx_2 \cdots dx_d = r^{d-1} dr \, d\Omega \tag{C.30}$$

where r is the hyperradius and where $d\Omega$ is the element of generalized solid angle. From the Hermiticity of the generalized angular momentum operator Λ^2, one can show that its eigenfunctions corresponding to different eigenvalues are orthogonal with respect to hyperangular integration. Thus from (C.27) it follows that

$$\int d\Omega \, h_{\lambda'}^* h_\lambda = 0 \quad \text{if } \lambda' \neq \lambda \tag{C.31}$$

In the particular case where $\lambda' = 0$, this becomes

$$\int d\Omega \; h_0^* h_\lambda = h_0^* \int d\Omega \; h_\lambda = 0 \quad \text{if} \; \lambda \neq 0 \qquad \text{(C.32)}$$

since the constant h_0^* can be factored out of the integration over generalized solid angle. Thus we obtain the important result:

$$\int d\Omega \; h_\lambda = 0 \quad \text{if} \; \lambda \neq 0 \qquad \text{(C.33)}$$

Let us now combine this result with equation (C.11), which shows the form of the canonical decomposition of a homogeneous polynomial f_n. From (C.11) and (C.33) it follows that if a homogeneous polynomial is integrated over generalized solid angle, the only term that will survive is the constant term in the canonical decomposition, i.e. h_0. But this term, together with the power of the hyperradius multiplying it, can be factored out of the integration. Thus we obtain the powerful angular and hyperangular integration theorem:

$$\int d\Omega \; f_n = \begin{cases} r^n h_0 \int d\Omega & n = \text{even} \\[2mm] 0 & n = \text{odd} \end{cases} \qquad \text{(C.34)}$$

where we have used the fact that when n is odd, the constant term h_0 does not occur in the canonical decomposition. We already have an explicit expression for h_0, namely equation (C.16). The only task remaining is to evaluate the total generalized solid angle, $\int d\Omega$. We can do this by comparing the integral of e^{-r^2} over the whole d-dimensional space when performed in Cartesian coordinates with the same integral performed in generalized spherical polar coordinates. Since the result must be the same, independent of the coordinate system used, we have:

$$\int_0^\infty dr \; r^{d-1} e^{-r^2} \int d\Omega = \prod_{j=1}^d \int_{-\infty}^\infty dx_j \; e^{-x_j^2} \qquad \text{(C.35)}$$

The hyperradial integral can be expressed in terms of the gamma function:

$$\int_0^\infty dr \; r^{d-1} e^{-r^2} = \frac{\Gamma(d/2)}{2} \qquad \text{(C.36)}$$

as can the integral over each of the Cartesian coordinates:

$$\int_{-\infty}^\infty dx_j \; e^{-x_j^2} = \Gamma(1/2) = \pi^{\frac{1}{2}} \qquad \text{(C.37)}$$

Inserting these results into (C.35) and solving for $\int d\Omega$, we obtain:

$$\int d\Omega = \frac{2\pi^{\frac{d}{2}}}{\Gamma\left(\frac{d}{2}\right)} \tag{C.38}$$

Finally, combining (C.11), (C.34) and (C.38), we have an explicit expression for the integral over generalized solid angle of any homogeneous polynomial of degree n:

$$\int d\Omega \ f_n = \begin{cases} \dfrac{2\pi^{d/2} r^n (d-2)!!}{\Gamma(d/2) n!! (d+n-2)!!} \Delta^{\frac{1}{2}n} f_n & n = \text{even} \\[4mm] 0 & n = \text{odd} \end{cases} \tag{C.39}$$

Now suppose that $F(\mathbf{x})$ is any function whatever that can be expanded in a convergent series of homogeneous polynomials. If the series has the form:

$$F(\mathbf{x}) = \sum_{n=0}^{\infty} f_n(\mathbf{x}) \tag{C.40}$$

then it follows from (C.39) that

$$\int d\Omega \ F(\mathbf{x}) = \frac{(d-2)!! 2\pi^{d/2}}{\Gamma\left(\frac{d}{2}\right)} \sum_{n=0,2,..}^{\infty} \frac{r^n}{n!!(n+d-2)!!} \Delta^{n/2} f_n(\mathbf{x}) \tag{C.41}$$

We can notice that at the point $\mathbf{x} = 0$, all terms in a polynomial vanish, except the constant term. Thus we have

$$\left\lfloor \Delta^{n/2} F(\mathbf{x}) \right\rfloor_{\mathbf{x}=0} = \Delta^{n/2} f_n(\mathbf{x}) \tag{C.42}$$

This allows us to rewrite (C.41) in the form

$$\int d\Omega \ F(\mathbf{x}) = \frac{(d-2)!! 2\pi^{d/2}}{\Gamma\left(\frac{d}{2}\right)} \sum_{\nu=0}^{\infty} \frac{r^{2\nu}}{(2\nu)!!(d+2\nu-2)!!} \left\lfloor \Delta^{\nu} F(\mathbf{x}) \right\rfloor_{\mathbf{x}=0} \tag{C.43}$$

where we have made the substitution $n = 2\nu$. In the case where $d = 3$, this reduces to

$$\int d\Omega \ F(\mathbf{x}) = 4\pi \sum_{\nu=0}^{\infty} \frac{r^{2\nu}}{(2\nu+1)!} \left\lfloor \Delta^{\nu} F(\mathbf{x}) \right\rfloor_{\mathbf{x}=0} \tag{C.44}$$

C.6   An alternative method for angular and hyperangular integrations

Theorem:

Let

$$I(\mathbf{n}) \equiv \int d\Omega \left(\frac{x_1}{r}\right)^{n_1} \left(\frac{x_2}{r}\right)^{n_2} \cdots \left(\frac{x_d}{r}\right)^{n_d} \qquad (C.45)$$

where $x_1, x_2, \ldots, x_d$ are the Cartesian coordinates of a d-dimensional space, $d\Omega$ is the generalized solid angle, r is the hyperradius, defined by

$$r^2 \equiv \sum_{j=1}^{d} x_j^2 \qquad (C.46)$$

and where the n_j's are positive integers or zero. Then

$$I(\mathbf{n}) = \begin{cases} \dfrac{\pi^{d/2}}{2^{(n/2-1)}\Gamma\left(\frac{d+n}{2}\right)} \displaystyle\prod_{j=1}^{d}(n_j - 1)!! & \text{if all the } n_j\text{'s are even} \\[2em] 0 & \text{otherwise} \end{cases}$$

$$\qquad (C.47)$$

where

$$n \equiv \sum_{j=1}^{d} n_j \qquad (C.48)$$

Proof:

We consider the integral

$$\int_0^\infty dr\, r^{d-1} e^{-r^2} \int d\Omega\, x_1^{n_1} x_2^{n_2} \cdots x_d^{n_d} = \prod_{j=1}^{d} \int_{-\infty}^{\infty} dx_j\, x_j^{n_j} e^{-x_j^2} \qquad (C.49)$$

If n_j is zero or a positive integer, then

$$\int_{-\infty}^{\infty} dx_j\, x_j^{n_j} e^{-x_j^2} = \begin{cases} \dfrac{(n_j - 1)!!\sqrt{\pi}}{2^{n_j/2}} & \text{if } n_j \text{ is even} \\[1.5em] 0 & \text{if } n_j \text{ is odd} \end{cases} \qquad (C.50)$$

so that the right-hand side of (C.49) becomes

$$\prod_{j=1}^{d} \int_{-\infty}^{\infty} dx_j\, x_j^{n_j} e^{-x_j^2} = \begin{cases} \dfrac{\pi^{d/2}}{2^{n/2}} \displaystyle\prod_{j=1}^{d}(n_j - 1)!! & \text{if all the } n_j \text{ are even} \\[2em] 0 & \text{otherwise} \end{cases}$$

$$\qquad (C.51)$$

The left-hand side of (5) can be written in the form

$$\int_0^\infty dr\ r^{d+n-1}e^{-r^2} \int d\Omega\ \left(\frac{x_1}{r}\right)^{n_1}\left(\frac{x_2}{r}\right)^{n_2}\cdots\left(\frac{x_d}{r}\right)^{n_d} = \frac{I(\mathbf{n})}{2}\Gamma\left(\frac{d+n}{2}\right)$$

(C.52)

Substituting (C.51) and (C.52) into (C.49), we obtain:

$$I(\mathbf{n}) = \begin{cases} \dfrac{\pi^{d/2}}{2^{(n/2-1)}\Gamma\left(\frac{d+n}{2}\right)} \displaystyle\prod_{j=1}^{d}(n_j-1)!! & \text{if all the } n_j \text{ are even} \\[3ex] 0 & \text{otherwise} \end{cases}$$

(C.53)

Q.E.D.

Comments:

In the special case where d=3, equation (C.47) becomes

$$\int d\Omega\ \left(\frac{x_1}{r}\right)^{n_1}\left(\frac{x_2}{r}\right)^{n_2}\left(\frac{x_d}{r}\right)^{n_3} = \begin{cases} \dfrac{4\pi}{(n+1)!!} \displaystyle\prod_{j=1}^{3}(n_j-1)!! & \text{all } n_j \text{ are even} \\[3ex] 0 & \text{otherwise} \end{cases}$$

(C.54)

Let us now consider a general polynomial (not necessarily homogeneous) of the form:

$$P(\mathbf{x}) = \sum_{\mathbf{n}} c_{\mathbf{n}}\ x_1^{n_1}x_2^{n_2}\cdots x_d^{n_d}$$

(C.55)

Then we have

$$\int d\Omega\ P(\mathbf{x}) = \sum_{\mathbf{n}} c_{\mathbf{n}} \int d\Omega\ x_1^{n_1}x_2^{n_2}\cdots x_d^{n_d} = \sum_{\mathbf{n}} c_{\mathbf{n}}\ r^n I(\mathbf{n})$$

(C.56)

It can be seen that equation (C.47) can be used to evaluate the generalized angular integral of any polynomial whatever, regardless of whether or not it is homogeneous.

It is interesting to ask what happens if the n_j's are not required to be zero or positive integers. If all the n_j's are real numbers greater than -1, then the right-hand side of (C.49) can still be evaluated and it has the form

$$\prod_{j=1}^{d} \int_{-\infty}^{\infty} dx_j\ x_j^{n_j}e^{-x_j^2} = \prod_{j=1}^{d} \frac{1}{2}\left(1+e^{i\pi n_j}\right)\Gamma\left(\frac{n_j+1}{2}\right)$$

(C.57)

Thus (C.47) becomes

$$I(\mathbf{n}) = \frac{2}{\Gamma\left(\frac{d+n}{2}\right)} \prod_{j=1}^{d} \frac{1}{2} \left(1 + e^{i\pi n_j}\right) \Gamma\left(\frac{n_j + 1}{2}\right) \qquad n_j > -1, \quad j = 1, \ldots, d$$

(C.58)

This more general equation reduces to (C.47) in the special case where the n_j's are required to be either zero or positive integers.

C.7 Angular integrations by a vector-pairing method

Let us consider the following integral in a 3-dimensional space:

$$I = \frac{1}{4\pi} \int d\Omega \; (\hat{\mathbf{x}} \cdot \hat{\mathbf{A}})(\hat{\mathbf{x}} \cdot \hat{\mathbf{B}})$$

(C.59)

where $\hat{\mathbf{A}}$ and $\hat{\mathbf{B}}$ are unit vectors. Since the integral I must be independent of $\mathbf{x}$ and invariant under rotations, it must be proportional to the scalar product, $\hat{\mathbf{A}} \cdot \hat{\mathbf{B}}$, which is the only scalar that can be made out of two vectors. The constant of proportionality can be found by considering the case where $\hat{\mathbf{A}} = \hat{\mathbf{B}}$, and in this way, one finds that

$$\frac{1}{4\pi} \int d\Omega \; (\hat{\mathbf{x}} \cdot \hat{\mathbf{A}})(\hat{\mathbf{x}} \cdot \hat{\mathbf{B}}) = \frac{1}{3}(\hat{\mathbf{A}} \cdot \hat{\mathbf{B}})$$

(C.60)

Building on this approach to angular integration, Avery, Ørmen and Michels [Avery and Ørmen, 1980], [Michels, 1981] were able to show that

$$\frac{1}{4\pi} \int d\Omega \; \prod_{j=1}^{N} (\hat{\mathbf{x}} \cdot \hat{\mathbf{A}}_j)^{n_j}$$

$$= \begin{cases} \dfrac{1}{(n+1)!!} \displaystyle\sum_{\lambda^*} \left(\prod_{k=1}^{N} \dfrac{n_k!}{(2\lambda_{kk})!!} \right) \prod_{i=1}^{j-1} \prod_{j=1}^{N} \dfrac{1}{\lambda_{ij}!} (\hat{\mathbf{A}}_i \cdot \hat{\mathbf{A}}_j)^{\lambda_{ij}} & n \text{ even} \\[20pt] 0 & n \text{ odd} \end{cases}$$

(C.61)

where

$$n \equiv n_1 + n_2 + n_3 + \cdots + n_N$$

(C.62)

In (C.61) the sum $\sum_{\lambda^*}$ denotes a sum over all sets of λ_{ij} values which are positive integers or zero and which fulfil the criteria

$$2\lambda_{jj} + \sum_{i=1}^{j-1} \lambda_{ij} + \sum_{i=j+1}^{N} \lambda_{ji} = n_j \qquad j = 1, 2, \ldots, N$$

(C.63)

For example, when $N = 2$, $n_1 = 3$ and $n_2 = 3$, the set of λ_{ij} values

$$\lambda_{11} = 1 \qquad \lambda_{22} = 1 \qquad \lambda_{12} = 1 \qquad\qquad (C.64)$$

fulfils (C.63), and likewise

$$\lambda_{11} = 0 \qquad \lambda_{22} = 0 \qquad \lambda_{12} = 3 \qquad\qquad (C.65)$$

satisfies (C.63). These are the only possibilities, and thus the sum in (C.61) contains two terms.

It is easy to extend these methods to spaces of higher dimension, and the relevant formulae can be found in references [Avery and Ørmen, 1980] and [Michels, 1981]. It is also possible to use (C.61) to evaluate integrals of the type

$$W_{l,l',l''} \equiv \frac{1}{4\pi} \int d\Omega \; P_l(\hat{\mathbf{x}} \cdot \hat{\mathbf{A}}) P_{l'}(\hat{\mathbf{x}} \cdot \hat{\mathbf{B}}) P_{l''}(\hat{\mathbf{x}} \cdot \hat{\mathbf{C}}) \qquad\qquad (C.66)$$

where P_l is a Legendre polynomial, and some examples are shown in the following table, where only nonzero values are shown. In order for $W_{l,l',l''}$ to be nonzero, $l + l' + l''$ must be even and $|l - l'| \leq l'' \leq l + l'$.

Table C.1 Integrals of products of Legendre polynomials, evaluated by the vector pairing method

(l, l', l'')	$W_{l,l',l''} \equiv \dfrac{1}{4\pi} \displaystyle\int d\Omega\ P_l(\hat{\mathbf{x}} \cdot \hat{\mathbf{A}}) P_{l'}(\hat{\mathbf{x}} \cdot \hat{\mathbf{B}}) P_{l''}(\hat{\mathbf{x}} \cdot \hat{\mathbf{C}})$
$(0,0,0)$	1
$(1,1,0)$	$\dfrac{1}{3}(\hat{\mathbf{A}} \cdot \hat{\mathbf{B}})$
$(1,1,2)$	$\dfrac{1}{15}\left[-(\hat{\mathbf{A}} \cdot \hat{\mathbf{B}}) + 3(\hat{\mathbf{A}} \cdot \hat{\mathbf{C}})(\hat{\mathbf{B}} \cdot \hat{\mathbf{C}}) \right]$
$(2,2,0)$	$\dfrac{1}{10}\left[-1 + 3(\hat{\mathbf{A}} \cdot \hat{\mathbf{B}})^2 \right]$
$(2,2,2)$	$\dfrac{1}{35}\left[2 - 3(\hat{\mathbf{A}} \cdot \hat{\mathbf{B}})^2 - 3(\hat{\mathbf{A}} \cdot \hat{\mathbf{B}})^2 - 3(\hat{\mathbf{A}} \cdot \hat{\mathbf{B}})^2 + 9(\hat{\mathbf{A}} \cdot \hat{\mathbf{B}})(\hat{\mathbf{A}} \cdot \hat{\mathbf{C}})(\hat{\mathbf{B}} \cdot \hat{\mathbf{C}}) \right]$
$(1,2,3)$	$\dfrac{3}{70}\left[-(\hat{\mathbf{A}} \cdot \hat{\mathbf{C}}) - 2(\hat{\mathbf{A}} \cdot \hat{\mathbf{B}})(\hat{\mathbf{B}} \cdot \hat{\mathbf{C}}) + 5(\hat{\mathbf{A}} \cdot \hat{\mathbf{C}})(\hat{\mathbf{B}} \cdot \hat{\mathbf{C}})^2 \right]$
$(3,3,0)$	$\dfrac{1}{14}\left[-3(\hat{\mathbf{A}} \cdot \hat{\mathbf{B}}) + 5(\hat{\mathbf{A}} \cdot \hat{\mathbf{B}})^3 \right]$

Appendix D

INTERELECTRON REPULSION INTEGRALS

D.1 The generalized Slater-Condon rules

The interelectron repulsion operator has the form

$$V' = \sum_{i=1}^{N} \sum_{j>i}^{N} \frac{1}{r_{ij}} \tag{D.1}$$

and it is a typical two-electron operator. Suppose that we would like to calculate the matrix element of V' between two configurations, each of which is represented by a Slater determinant:

$$|\Phi_\nu\rangle = |f_1 f_2 f_3 \cdots| \tag{D.2}$$

$$|\Phi_{\nu'}\rangle = |g_1 g_2 g_3 \cdots| \tag{D.3}$$

If we cannot assume orthonormality between the orbitals of the two configurations, we must make use of the generalized Slater-Condon rules [Rettrup, 2003]. The first of these rules states that

$$\langle \Phi_\nu | \Phi_{\nu'} \rangle = |S| \tag{D.4}$$

where S is the matrix of overlap integrals:

$$S \equiv \begin{pmatrix} \langle f_1|g_1\rangle \ \langle f_1|g_2\rangle \ \cdots \ \langle f_1|g_N\rangle \\ \langle f_2|g_1\rangle \ \langle f_2|g_2\rangle \ \cdots \ \langle f_2|g_N\rangle \\ \vdots \qquad \vdots \qquad\quad \vdots \\ \langle f_3|g_1\rangle \ \langle f_3|g_2\rangle \ \cdots \ \langle f_3|g_N\rangle \end{pmatrix} \tag{D.5}$$

and where $|S|$ is its determinant. Using the generalized Slater-Condon rule for two-electron operators, we can write the matrix element V' between two configurations in the form

$$\langle \Phi_\nu | V' | \Phi_{\nu'} \rangle = \sum_{i=1}^{N} \sum_{j=i+1}^{N} \sum_{k=1}^{N} \sum_{l=k+1}^{N} (-1)^{i+j+k+l} C_{ij;kl} |S_{ij;kl}| \qquad \text{(D.6)}$$

where

$$C_{ij;kl} \equiv \int d\tau_1 \int d\tau_2 \; f_i^*(1) f_j^*(2) \frac{1}{r_{12}} \left[g_k(1) g_l(2) - g_l(1) g_k(2) \right] \qquad \text{(D.7)}$$

and where $|S_{ij;kl}|$ is the determinant of the matrix formed from S by deleting the ith and jth rows and the kth and lth columns. For the sake of completeness, we should mention the generalized Slater-Condon rule appropriate for one-electron operators, i.e. operators of the form

$$V = v(1) + v(2) + v(3) + \cdots + v(N) \qquad \text{(D.8)}$$

In that case the rule is

$$\langle \Phi_\nu | V | \Phi_{\nu'} \rangle = \sum_{i=1}^{N} \sum_{j=1}^{N} (-1)^{i+j} \langle f_i | v | g_j \rangle |S_{ij}| \qquad \text{(D.9)}$$

where $|S_{ij}|$ is the determinant of the matrix obtained from S by deleting the ith row and the jth column.

D.2 Separation of atomic integrals into radial and angular parts

For the case of atoms, the problem of evaluating the interelectron repulsion integrals shown in equation (D.7) can be solved provided that we are able to evaluate integrals of the form

$$J = \int_0^\infty dr_1 \; r_1^{2+j_1} e^{-\zeta_1 r_1} \int_0^\infty dr_2 \; r_2^{2+j_2} e^{-\zeta_2 r_2}$$
$$\times \int d\Omega_1 \; W_1(\hat{\mathbf{x}}_1) \int d\Omega_2 \; W_2(\hat{\mathbf{x}}_2) \frac{1}{r_{12}} \qquad \text{(D.10)}$$

where $W_1(\hat{\mathbf{x}}_1)$ and $W_2(\hat{\mathbf{x}}_2)$ are angular functions. We can separate this integral into radial and angular parts by introducing the expansion:

$$\frac{1}{r_{12}} \equiv \frac{1}{|\mathbf{x}_1 - \mathbf{x}_2|} = \sum_{l=0}^{\infty} \frac{r_<^l}{r_>^{l+1}} P_l(\hat{\mathbf{x}}_1 \cdot \hat{\mathbf{x}}_2) \qquad \text{(D.11)}$$

from which we obtain

$$J = \sum_{l=0}^{\infty} a_l I_{l,j_1,j_2}(\zeta_1, \zeta_2) \tag{D.12}$$

where

$$a_l \equiv \int d\Omega_1 \ W_1(\hat{\mathbf{x}}_1) \int d\Omega_2 \ W_2(\hat{\mathbf{x}}_2) P_l(\hat{\mathbf{x}}_1 \cdot \hat{\mathbf{x}}_2) \tag{D.13}$$

and

$$I_{l,j_1,j_2}(\zeta_1, \zeta_2) \equiv \int_0^{\infty} dr_1 \ r_1^{j_1+2} e^{-\zeta_1 r_1} \int_0^{\infty} dr_2 \ r_2^{j_2+2} e^{-\zeta_2 r_2} \frac{r_<^l}{r_>^{l+1}} \tag{D.14}$$

The sum in (D.12) is written as an infinite sum, but it contains only a few terms, since only a few of the coefficients a_l are nonzero. These coefficients may be evaluated using the angular integration methods discussed in Appendix C.

D.3 Evaluation of the radial integrals in terms of hypergeometric functions

The result of the first radial integration

$$\int_0^{\infty} dr_2 \ r_2^{j_2+2} e^{-\zeta_2 r_2} \frac{r_<^l}{r_>^{l+1}}$$
$$= \frac{1}{r_1^{l+1}} \int_0^{r_1} dr_2 \ r_2^{j_2+l+2} e^{-\zeta_2 r_2} + r_1^l \int_{r_1}^{\infty} dr_2 \ r_2^{j_2-l+1} e^{-\zeta_2 r_2} \tag{D.15}$$

can be expressed in terms of incomplete gamma functions, since

$$\frac{1}{r_1^{l+1}} \int_0^{r_1} dr_2 \ r_2^{j_2+l+2} e^{-\zeta_2 r_2} = \frac{1}{(\zeta_2 r_1)^{l+1} \zeta_2^{j_2+2}} \ \gamma(j_2 + l + 3, \zeta_2 r_1) \tag{D.16}$$

and

$$r_1^l \int_{r_1}^{\infty} dr_2 \ r_2^{j_2-l+1} e^{-\zeta_2 r_2} = \frac{(\zeta_2 r_1)^l}{\zeta_2^{j_2+2}} \ \Gamma(j_2 - l + 2, \zeta_2 r_1) \tag{D.17}$$

where

$$\gamma(a, z) \equiv \int_0^z dt \ t^{a-1} e^{-t} \tag{D.18}$$

and

$$\Gamma(a, z) \equiv \int_z^{\infty} dt \ t^{a-1} e^{-t} \tag{D.19}$$

By inserting this result into equation (D.14) and consulting (for example) [Gradshteyn and Ryshik, 1965], we finally obtain an expression for I_l in terms of hypergeometric functions:

$$
\begin{aligned}
I_{l,j_1,j_2}(\zeta_1,\zeta_2) &= \int_0^\infty dr_1 \, r_1^{j_1+2} e^{-\zeta_1 r_1} \int_0^\infty dr_2 \, r_2^{j_2+2} e^{-\zeta_2 r_2} \frac{r_<^l}{r_>^{l+1}} \\
&= \frac{\Gamma(j_1+j_2+5)}{(\zeta_1+\zeta_2)^{j_1+j_2+5}} \\
&\quad \times \left[\frac{{}_2F_1\left(1,j_1+j_2+5|j_2+l+4|\zeta_2/(\zeta_1+\zeta_2)\right)}{j_2+l+3} \right. \\
&\qquad \left. + \frac{{}_2F_1\left(1,j_1+j_2+5|j_1+l+4|\zeta_1/(\zeta_1+\zeta_2)\right)}{j_1+l+3} \right]
\end{aligned}
$$

$$(D.20)$$

A density composed of the product of two atomic orbitals has the form

$$
\begin{aligned}
\rho_{\mu_1,\mu_2}(\mathbf{x}_1) &= \chi^*_{n_1,l_1,m_1,Q_1}(\mathbf{x}_1)\chi_{n_1,l_2,m_2,Q_2}(\mathbf{x}_1) \\
&= R_{n_1,l_1,Q_1}(r_1)R_{n_2,l_2,Q_2}(r_1)Y^*_{l_1,m_1}(\hat{\mathbf{x}}_1)Y_{l_2,m_2}(\hat{\mathbf{x}}_1) \\
&\equiv W_1(\hat{\mathbf{x}}_1)\sum_{j_1} c_{j_1} r_1^{j_1} e^{-\zeta_1 r_1}
\end{aligned}
$$

$$(D.21)$$

Here

$$
W_1(\hat{\mathbf{x}}_1) \equiv Y^*_{l_1,m_1}(\hat{\mathbf{x}}_1)Y_{l_2,m_2}(\hat{\mathbf{x}}_1)
$$

$$(D.22)$$

while

$$
\sum_{j_1} c_{j_1} r_1^{j_1} e^{-\zeta_1 r_1} \equiv R_{n_1,l_1,Q_1}(r_1)R_{n_2,l_2,Q_2}(r_1)
$$

$$(D.23)$$

and

$$
\zeta_1 \equiv \frac{Q_1}{n_1} + \frac{Q_2}{n_2}
$$

$$(D.24)$$

Similarly we can write

$$
\rho_{\mu_3,\mu_4}(\mathbf{x}_2) \equiv W_2(\hat{\mathbf{x}}_2)\sum_{j_2} c_{j_2} r_2^{j_2} e^{-\zeta_2 r_2}
$$

$$(D.25)$$

where

$$
W_2(\hat{\mathbf{x}}_2) \equiv Y^*_{l_3,m_3}(\hat{\mathbf{x}}_2)Y_{l_4,m_4}(\hat{\mathbf{x}}_2)
$$

$$(D.26)$$

while

$$
\sum_{j_2} c_{j_2} r_2^{j_2} e^{-\zeta_2 r_2} \equiv R_{n_3,l_3,Q_3}(r_2)R_{n_4,l_4,Q_4}(r_2)
$$

$$(D.27)$$

and

$$\zeta_2 \equiv \frac{Q_3}{n_3} + \frac{Q_4}{n_4} \tag{D.28}$$

Thus the atomic interelectron repulsion integral between two densities, each of which is a product of two atomic orbitals, can be written as

$$J_{\mu_1,\mu_2;\mu_3,\mu_4} = \int d^3x_1 \int d^3x_2 \; \rho_{\mu_1,\mu_2}(\mathbf{x}_1) \frac{1}{|\mathbf{x}_1 - \mathbf{x}_2|} \rho_{\mu_3,\mu_4}(\mathbf{x}_2)$$

$$= \sum_l a_l \sum_{j_1,j_2} c_{j_1} c_{j_2} I_{l,j_1,j_2}(\zeta_1,\zeta_2) \tag{D.29}$$

where a_l and $I_{l,j_1,j_2}(\zeta_1,\zeta_2)$ are given respectively by equations (D.13) and (D.20).

D.4 Evaluation of the angular integrals in terms of Condon-Shortley coefficients

We now introduce the notation

$$\int d\Omega_1 \; Y_{l,m}(\hat{\mathbf{x}}_1) W_1(\hat{\mathbf{x}}_1) = \int d\Omega_1 \; Y_{l,m}(\hat{\mathbf{x}}_1) Y^*_{l_1,m_1}(\hat{\mathbf{x}}_1) Y_{l_2,m_2}(\hat{\mathbf{x}}_1)$$

$$= \delta_{m,m_1-m_2} \int d\Omega_1 \; Y_{l,m_1-m_2}(\hat{\mathbf{x}}_1) Y^*_{l_1,m_1}(\hat{\mathbf{x}}_1) Y_{l_2,m_2}(\hat{\mathbf{x}}_1)$$

$$\equiv \delta_{m,m_1-m_2} \sqrt{\frac{2l+1}{4\pi}} C^l_{l_1,m_1,l_2,m_2} \tag{D.30}$$

where $C^l_{l_1,m_1,l_2,m_2}$ is a Condon-Shortley coefficient. Similarly

$$\int d\Omega_2 \; Y^*_{l,m}(\hat{\mathbf{x}}_2) W_2(\hat{\mathbf{x}}_2) = \int d\Omega_2 \; Y^*_{l,m}(\hat{\mathbf{x}}_2) Y^*_{l_3,m_3}(\hat{\mathbf{x}}_2) Y_{l_4,m_4}(\hat{\mathbf{x}}_2)$$

$$= \delta_{m,m_4-m_3} \int d\Omega_2 \; Y^*_{l,m_4-m_3}(\hat{\mathbf{x}}_2) Y^*_{l_3,m_3}(\hat{\mathbf{x}}_2) Y_{l_4,m_4}(\hat{\mathbf{x}}_2)$$

$$\equiv \delta_{m,m_4-m_3} (-1)^{m_4-m_3} \sqrt{\frac{2l+1}{4\pi}} C^l_{l_3,m_3,l_4,m_4} \tag{D.31}$$

Then the angular integral, written in terms of Condon-Shortley coefficients, becomes:

$$a_l \equiv \int d\Omega_1 \; W_1(\hat{\mathbf{x}}_1) \int d\Omega_2 \; W_2(\hat{\mathbf{x}}_2) P_l(\hat{\mathbf{x}}_1 \cdot \hat{\mathbf{x}}_2)$$

$$= \frac{4\pi}{2l+1} \sum_{m=-l}^{l} \int d\Omega_1 \; Y_{l,m}(\hat{\mathbf{x}}_1) W_1(\hat{\mathbf{x}}_1) \int d\Omega_2 \; Y^*_{l,m}(\hat{\mathbf{x}}_2) W_2(\hat{\mathbf{x}}_2)$$

$$= \delta_{m_1-m_2,m_4-m_3} (-1)^{m_1-m_2} C^l_{l_1,m_1,l_2,m_2} C^l_{l_3,m_3,l_4,m_4} \tag{D.32}$$

Appendix E

GAUSSIAN EXPANSION OF MOLECULAR STURMIANS

E.1 Expansions of Coulomb Sturmian densities in terms of Gaussians

Let us suppose that we have found coefficients $\gamma_{n,i}$ such that

$$e^{-s} \approx \sum_i \gamma_{0,i} e^{-\alpha_i s^2} \tag{E.1}$$

$$se^{-s} \approx \sum_i \gamma_{1,i} e^{-\alpha_i s^2} \tag{E.2}$$

and

$$s^2 e^{-s} \approx \sum_i \gamma_{2,i} e^{-\alpha_i s^2} \tag{E.3}$$

(see Table 5.5). We now introduce the regular solid harmonics R_l^m defined by

$$\mathsf{R}_l^m(\mathbf{x}) \equiv \sqrt{\frac{4\pi}{2l+1}} r^l Y_{lm}(\hat{\mathbf{x}}) \tag{E.4}$$

(here, and throughout this book, a unit vector is indicated by a "hat"). The first few of these are

$$\begin{aligned}
\mathsf{R}_0^0(\mathbf{x}) &= 1 \\
\mathsf{R}_1^{-1}(\mathbf{x}) &= (x - iy)/\sqrt{2} \\
\mathsf{R}_1^0(\mathbf{x}) &= z \\
\mathsf{R}_1^1(\mathbf{x}) &= -(x + iy)/\sqrt{2}
\end{aligned} \tag{E.5}$$

Expressed in terms of the regular solid harmonics, the Coulomb Sturmian basis functions can be written as

$$
\begin{aligned}
\chi_{n,l,m}(\mathbf{x}) &= k^{3/2}\tilde{R}_{n,l}(s)Y_{l,m}(\hat{\mathbf{x}}) \\
&= k^{3/2}\sqrt{\frac{2l+1}{4\pi}}\tilde{R}_{n,l}(s)s^{-l}\mathrm{R}_l^m(k\mathbf{x}) \\
&\approx k^{3/2}\sum_i \Gamma_{n,l,i}e^{-\alpha_i|k\mathbf{x}|^2}\mathrm{R}_l^m(k\mathbf{x})
\end{aligned}
\tag{E.6}
$$

where $\tilde{R}_{n.l}(s) \equiv R_{n,l}(r)/k^{3/2}$, and where $s \equiv kr$. The coefficients $\Gamma_{n,l,i}$ are defined by the relationship

$$
\sqrt{\frac{2l+1}{4\pi}}\tilde{R}_{n,l}(s)s^{-l} \approx \sum_i \Gamma_{n,l,i}e^{-\alpha_i s^2}
\tag{E.7}
$$

Then

$$
\chi_{n,l,m}(\mathbf{x}-\mathbf{X}_a) \approx k^{3/2}\sum_i \Gamma_{n,l,i}\ e^{-\alpha_i|k\mathbf{x}-k\mathbf{X}_a|^2}\mathrm{R}_l^m(k\mathbf{x}-k\mathbf{X}_a)
\tag{E.8}
$$

We next let

$$
\mathbf{s} \equiv k\mathbf{x} \qquad\qquad \mathbf{S}_a \equiv k\mathbf{X}_a
\tag{E.9}
$$

and

$$
\begin{aligned}
f_{i,j}(\mathbf{s}) &= e^{-\alpha_i|\mathbf{s}-\mathbf{S}_a|^2-\alpha_j|\mathbf{s}-\mathbf{S}_{a'}|^2} \\
&= e^{-\frac{\alpha_i\alpha_j}{\alpha_i+\alpha_j}|\mathbf{S}_a-\mathbf{S}_{a'}|^2}e^{-(\alpha_i+\alpha_j)|\mathbf{s}-\mathbf{S}_{i,j}|^2}
\end{aligned}
\tag{E.10}
$$

where

$$
\mathbf{S}_{i,j} \equiv \frac{\alpha_i\mathbf{S}_a+\alpha_j\mathbf{S}_{a'}}{\alpha_i+\alpha_j}
\tag{E.11}
$$

Equation (E.10) is just a simple case of the Gaussian product theorem, which tells us that the product of two Gaussians is another Gaussian located at an intermediate position. Combining equations (E.9)-(E.11), we obtain the relationship

$$
\begin{aligned}
\rho_{\tau,\tau'}(\mathbf{x}) &= \chi_{n,l,m}^*(\mathbf{x}-\mathbf{X}_a)\chi_{n',l',m'}(\mathbf{x}-\mathbf{X}_{a'}) \\
&\approx k^3\mathrm{R}_l^m(\mathbf{s}-\mathbf{S}_a)^*\mathrm{R}_{l'}^{m'}(\mathbf{s}-\mathbf{S}_{a'}) \\
&\quad \times \sum_i\sum_j \Gamma_{n',l',j}\Gamma_{n,l,i}\ f_{i,j}(\mathbf{s})
\end{aligned}
\tag{E.12}
$$

The next step is to take the Fourier transform of (E.12):

$$
\begin{aligned}
\rho_{\tau,\tau'}^{t}(\mathbf{q}) &= \frac{k^3}{(2\pi)^{3/2}} \int d^3x \ e^{-i\mathbf{p}\cdot\mathbf{x}} \ R_l^m(\mathbf{s}-\mathbf{S}_a)^* R_{l'}^{m'}(\mathbf{s}-\mathbf{S}_{a'}) \\
&\quad \times \sum_i \sum_j \ \Gamma_{n',l',j}\Gamma_{n,l,i} \ f_{i,j}(\mathbf{s}) \\
&= \frac{1}{(2\pi)^{3/2}} \sum_i \sum_j \ \Gamma_{n',l',j}\Gamma_{n,l,i} \\
&\quad \times \int d^3s \ e^{-i\mathbf{q}\cdot\mathbf{s}} \ R_l^m(\mathbf{s}-\mathbf{S}_a)^* R_{l'}^{m'}(\mathbf{s}-\mathbf{S}_{a'}) \ f_{i,j}(\mathbf{s})
\end{aligned}
\tag{E.13}
$$

where we shift to the variables s and $\mathbf{q}$ defined by

$$
\begin{aligned}
\mathbf{q} &\equiv \frac{1}{k}\mathbf{p} \\
d^3s &\equiv k^3 \ d^3x
\end{aligned}
\tag{E.14}
$$

We now notice that from (E.10)

$$
\begin{aligned}
f_{i,j}^t(\mathbf{q}) &\equiv \frac{1}{(2\pi)^{3/2}} \int d^3s \ e^{-i\mathbf{q}\cdot\mathbf{s}} \ f_{i,j}(\mathbf{s}) \\
&= C_{i,j} e^{-\alpha_{i,j}q^2 - i\mathbf{q}\cdot\mathbf{S}_{i,j}}
\end{aligned}
\tag{E.15}
$$

with

$$
C_{i,j} \equiv \frac{1}{[2(\alpha_i + \alpha_j)]^{3/2}} \exp\left[-\frac{\alpha_i\alpha_j}{\alpha_i + \alpha_j}|\mathbf{S}_a - \mathbf{S}_b|^2\right]
\tag{E.16}
$$

and

$$
\alpha_{i,j} \equiv \frac{1}{4(\alpha_i + \alpha_j)} \qquad \mathbf{S}_{i,j} \equiv \frac{\alpha_i\mathbf{S}_a + \alpha_j\mathbf{S}_{a'}}{\alpha_i + \alpha_j}
\tag{E.17}
$$

If we can construct an operator $R_l^m(i\nabla_q - \mathbf{S}_a)$ such that

$$
R_l^m(i\nabla_q - \mathbf{S}_a)e^{-i\mathbf{q}\cdot\mathbf{s}} = R_l^m(\mathbf{s}-\mathbf{S}_a)e^{-i\mathbf{q}\cdot\mathbf{s}}
\tag{E.18}
$$

then we will have

$$
\begin{aligned}
&\frac{1}{(2\pi)^{3/2}} \int d^3s \ e^{-i\mathbf{q}\cdot\mathbf{s}} \ R_l^m(\mathbf{s}-\mathbf{S}_a)^* R_{l'}^{m'}(\mathbf{s}-\mathbf{S}_{a'}) \ f_{i,j}(\mathbf{s}) \\
&= R_l^m(i\nabla_q - \mathbf{S}_a)^* R_{l'}^{m'}(i\nabla_q - \mathbf{S}_{a'})f_{i,j}^t(\mathbf{q}) \\
&\equiv P_{l,m;l'm'}^{i,j}(\mathbf{q})f_{i,j}^t(\mathbf{q})
\end{aligned}
\tag{E.19}
$$

where $P_{l,m;l'm'}^{i,j}(\mathbf{q})$ is a polynomial. We can indeed construct such an operator from the displaced solid harmonic $R_l^m(\mathbf{s}-\mathbf{S}_a)$ if we replace $\mathbf{s}$ everywhere by $i\nabla_q$ where ∇_q is the gradient operator with respect to $\mathbf{q}$. Then, combining (E.19) with (E.13), we obtain

$$\rho_{\tau_1,\tau_2}^t(\mathbf{q}) = \sum_{i_1,i_2} \Gamma_{n_1,l_1,i_1}\Gamma_{n_2,l_2,i_2} P_{l_1,m_1;l_2,m_2}^{i_1,i_2}(\mathbf{q})f_{i_1,i_2}^t(\mathbf{q}) \qquad (E.20)$$

We now remember that

$$
\begin{aligned}
J_{\tau_1,\tau_2;\tau_3,\tau_4} &= 4\pi \int d^3p \frac{1}{p^2} \rho_{\tau_1,\tau_2}^t(\mathbf{p})\rho_{\tau_3,\tau_4}^t(-\mathbf{p}) \\
&= 4\pi \int_0^\infty dp \int d\Omega_p \ \rho_{\tau_1,\tau_2}^t(\mathbf{p})\rho_{\tau_3,\tau_4}^t(-\mathbf{p}) \qquad (E.21) \\
&= 4\pi k \int_0^\infty dq \int d\Omega_q \ \rho_{\tau_1,\tau_2}^t(\mathbf{q})\rho_{\tau_3,\tau_4}^t(-\mathbf{q})
\end{aligned}
$$

and thus

$$
\begin{aligned}
J_{\tau_1,\tau_2;\tau_3,\tau_4} &= 4\pi k \int_0^\infty dq \int d\Omega_q \ \rho_{\tau_1,\tau_2}^t(\mathbf{q})\rho_{\tau_3,\tau_4}^t(-\mathbf{q}) \\
&= 4\pi k \sum_{\mathbf{i}} \Gamma_{n_1,l_1,i_1}\Gamma_{n_2,l_2,i_2}\Gamma_{n_3,l_3,i_3}\Gamma_{n_4,l_4,i_4} \int_0^\infty dq \int d\Omega_q \\
&\quad \times P_{l_1,m_1;l_2,m_2}^{i_1,i_2}(\mathbf{q}) \ P_{l_3,m_3;l_4,m_4}^{i_3,i_4}(-\mathbf{q})f_{i_1,i_2}^t(\mathbf{q})f_{i_3,i_4}^t(-\mathbf{q})
\end{aligned}
$$
$$(E.22)$$

Then, combining (E.22) and (E.15), we can write:

$$
\begin{aligned}
J_{\tau_1,\tau_2;\tau_3,\tau_4} =&\, 4\pi k \sum_{\mathbf{i}} \Gamma_{n_1,l_1,i_1}\Gamma_{n_2,l_2,i_2}\Gamma_{n_3,l_3,i_3}\Gamma_{n_4,l,i_4} \\
&\times \int_0^\infty dq \ e^{-\alpha_{\mathbf{i}} q^2} \int d\Omega_q \ P_{\mathbf{l},\mathbf{m}}^{\mathbf{i}}(\mathbf{q})e^{i\mathbf{q}\cdot\mathbf{S_i}}
\end{aligned}
$$
$$(E.23)$$

where $\sum_{\mathbf{i}} \equiv \sum_{i_1,i_2,i_3,i_4}$ and where

$$
\begin{aligned}
\mathbf{S_i} &\equiv \mathbf{S}_{i_3,i_4} - \mathbf{S}_{i_1,i_2} \\
P_{\mathbf{l},\mathbf{m}}^{\mathbf{i}}(\mathbf{q}) &\equiv P_{l_1,m_1;l_1,m_2}^{i_1,i_2}(\mathbf{q}) \ P_{l_3,m_3;l_4,m_4}^{i_3,i_4}(-\mathbf{q}) \\
\alpha_{\mathbf{i}} &\equiv \frac{1}{4(\alpha_{i_1}+\alpha_{i_2})} + \frac{1}{4(\alpha_{i_3}+\alpha_{i_4})}
\end{aligned}
$$
$$(E.24)$$

The integral in (E.23) can be rewritten as

$$
\begin{aligned}
I_{\mathbf{l},\mathbf{m}}^{\mathbf{i}}(\mathbf{S_i}) &\equiv \int_0^\infty dq \ e^{-\alpha_{\mathbf{i}} q^2} \int d\Omega_q \ P_{\mathbf{l},\mathbf{m}}^{\mathbf{i}}(\mathbf{q}) \ e^{i\mathbf{p}\cdot\mathbf{S_i}} \\
&= 4\pi \sum_l i^l \sum_{m=-l}^l Y_{l,m}(\hat{\mathbf{S}}_{\mathbf{i}}) \int_0^\infty dq \ e^{-\alpha_{\mathbf{i}} q^2} j_l(qS_i) \qquad (E.25) \\
&\quad \times \int d\Omega_q \ Y_{l,m}^*(\hat{\mathbf{q}})P_{\mathbf{l},\mathbf{m}}^{\mathbf{i}}(\mathbf{q})
\end{aligned}
$$

where we have made use of the expansion of a plane wave in terms of spherical harmonics and spherical Bessel functions. The resulting radial integrals can be evaluated exactly by Mathematica. For example,

$$\int_0^\infty dq\ e^{-\alpha_i q^2} j_0(qS_i) = \frac{\pi}{2S_i}\text{Erf}\left[\frac{S_i}{2\sqrt{\alpha_i}}\right] \tag{E.26}$$

where Erf is the error function, while

$$\int_0^\infty dq\ q\ e^{-\alpha_i q^2} j_1(qS_i) = \frac{\pi}{2S_i^2}\text{Erf}\left[\frac{S_i}{2\sqrt{\alpha_i}}\right] - \sqrt{\frac{\pi}{\alpha_i}}\frac{1}{S_i}\text{Exp}\left[-\frac{S_i^2}{4a_i}\right] \tag{E.27}$$

Thus we finally obtain

$$J_{\tau_1,\tau_2;\tau_3,\tau_4} = 4\pi k \sum_i \Gamma_{n_1,l_1,i_1}\Gamma_{n_2,l_2,i_2}\Gamma_{n_3,l_3,i_3}\Gamma_{n_4,l_4,i_4}I_{l,m}^i(\mathbf{S_i}) \tag{E.28}$$

Notice that $J_{\tau_1,\tau_2;\tau_3,\tau_4}/k$ is independent of k. The integrals can thus be evaluated once and for all and stored. They scale automatically with scaling of the basis set.

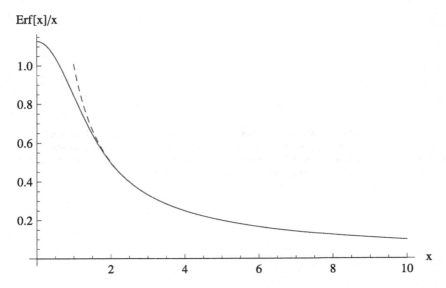

Fig. E.1 *The solid line in this figure shows the function* Erf[x]/x *while the dotted line compares it with 1/x. Since* Erf[x] *approaches 1 as x increases,* Erf[x]/x *approaches 1/x.*

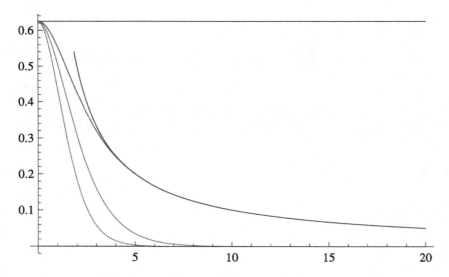

Fig. E.2   *Interelectron repulsion integrals $J_{\tau_1,\tau_2,\tau_3,\tau_4}/k$ for $\mathbf{n} = (1,1,1,1)$ and $\mathbf{l} = (0,0,0,0)$ as functions of S for the $(S,S,0,0)$, $(0,S,S,S)$ and $(0,S,0,S)$ cases. The $(S,S,0,0)$ case is compared with the curve $1/S$, which it approaches with increasing S. At the point $S = 0$, all the curves give the atomic value, $5/8$.*

Appendix F

EXPANSION OF DISPLACED FUNCTIONS IN TERMS OF LEGENDRE POLYNOMIALS

F.1 Displaced spherically symmetric functions

When a function F depends only on the distance $|\mathbf{x}_1 - \mathbf{x}_2|$, it is always possible to make an expansion of the form:

$$F(|\mathbf{x}_1 - \mathbf{x}_2|) = \sum_{l=0}^{\infty} f_l(r_1, r_2) P_l(\hat{\mathbf{x}}_1 \cdot \hat{\mathbf{x}}_2) \tag{F.1}$$

We would like to calculate the functions $f_l(r_1, r_2)$. To do this, we multiply (F.1) from the left by $P_{l'}(\hat{\mathbf{x}}_1 \cdot \hat{\mathbf{x}}_2)$ and integrate over solid angle, we obtain:

$$\int d\Omega \, P_{l'}(\hat{\mathbf{x}}_1 \cdot \hat{\mathbf{x}}_2) F(|\mathbf{x}_1 - \mathbf{x}_2|) = \sum_{l=0}^{\infty} f_l(r_1, r_2) \int d\Omega \, P_{l'}(\hat{\mathbf{x}}_1 \cdot \hat{\mathbf{x}}_2) P_l(\hat{\mathbf{x}}_1 \cdot \hat{\mathbf{x}}_2) \tag{F.2}$$

The angular integral on the right-hand side of (F.2) can be performed using the orthogonality of the Legendre polynomials:

$$\int d\Omega \, P_{l'}(\hat{\mathbf{x}}_1 \cdot \hat{\mathbf{x}}_2) P_l(\hat{\mathbf{x}}_1 \cdot \hat{\mathbf{x}}_2) = 2\pi \int_0^{\pi} d\theta \, \sin\theta P_{l'}(\cos\theta) P_l(\cos\theta)$$

$$= \frac{4\pi}{2l+1} \delta_{l',l} \tag{F.3}$$

and we obtain

$$f_l(r_1, r_2) = \frac{2l+1}{4\pi} \int d\Omega \, P_l(\hat{\mathbf{x}}_1 \cdot \hat{\mathbf{x}}_2) F(|\mathbf{x}_1 - \mathbf{x}_2|) \tag{F.4}$$

Remembering that

$$|\mathbf{x}_1 - \mathbf{x}_2| = \left(r_1^2 + r_2^2 - 2r_1 r_2 \cos\theta\right)^{1/2} \tag{F.5}$$

where

$$\cos\theta \equiv \hat{\mathbf{x}}_1 \cdot \hat{\mathbf{x}}_2 \tag{F.6}$$

we can rewrite (F.4) in the form

$$f_l(r_1, r_2) = \frac{2l+1}{2} \int_0^{\pi} d\theta \, \sin\theta P_l(\cos\theta) F \left(\left(r_1^2 + r_2^2 - 2r_1 r_2 \cos\theta \right)^{\frac{1}{2}} \right) \quad \text{(F.7)}$$

The integral in (F.7) can be evaluated directly, either numerically or else by expanding $F \left((r_1^2 + r_2^2 - 2r_1 r_2 \cos\theta)^{1/2} \right)$ as a Taylor series in $\cos\theta$.

F.2   An alternative method

For some types of potentials it may be convenient to use an alternative method first introduced by Peter Sommer-Larsen [Sommer-Larsen, 1990]. In this second method, we let the variable r be defined by

$$r \equiv \left(r_1^2 + r_2^2 - 2r_1 r_2 \cos\theta \right)^{1/2} \quad \text{(F.8)}$$

Then

$$\cos\theta = \frac{r_1^2 + r_2^2 - r^2}{2r_1 r_2} \quad \text{(F.9)}$$

and

$$\sin\theta \, d\theta = \frac{r}{r_1 r_2} \, dr \quad \text{(F.10)}$$

while equation (F.7) becomes

$$f_l(r_1, r_2) = \frac{2l+1}{2r_1 r_2} \int_{|r_1-r_2|}^{r_1+r_2} dr \, rF(r) P_l \left(\frac{r_1^2 + r_2^2 - r^2}{2r_1 r_2} \right) \quad \text{(F.11)}$$

Often the required integrals can be expressed in terms of incomplete Γ-functions. For example when

$$F(r) = r^n e^{-\zeta r} \quad \text{(F.12)}$$

we have

$$f_l(r_1, r_2) = \frac{2l+1}{2r_1 r_2} \int_{|r_1-r_2|}^{r_1+r_2} dr \, r^{n+1} e^{-\zeta r} P_l \left(\frac{r_1^2 + r_2^2 - r^2}{2r_1 r_2} \right) \quad \text{(F.13)}$$

Integrals of this type can be evaluated exactly by Mathematica for $l = 0, 1, 2, 3, \ldots$ and $n \geq -1$. The result is

$$f_0(r_1, r_2) = \frac{1}{2\zeta^{n+2} r_1 r_2} \left(\Gamma[n+2, \zeta|r_1 - r_2|] - \Gamma[n+2, \zeta(r_1 + r_2)] \right) \quad \text{(F.14)}$$

$$f_1(r_1, r_2)$$
$$= \frac{3}{4\zeta^{n+4} r_1^2 r_2^2} \left(\zeta^2(r_1^2 + r_2^2)\{\Gamma[n+2, \zeta|r_1 - r_2|] - \Gamma[n+2, \zeta(r_1 + r_2)]\} \right.$$
$$\left. - \Gamma[n+4, \zeta|r_1 - r_2|] + \Gamma[n+4, \zeta(r_1 + r_2)] \right) \quad \text{(F.15)}$$

and so on, where the incomplete Γ-functions are defined by

$$\Gamma[a, z] \equiv \int_z^{\infty} dt \, t^{a-1} e^{-t} \quad \text{(F.16)}$$

F.3   A screened Coulomb potential

For example, suppose that $n = -1$, so that

$$F(r) = \frac{e^{-\zeta r}}{r} \tag{F.17}$$

and $l = 0$. Then (F.11) yields

$$f_0(r_1, r_2) = \frac{1}{2r_1 r_2} \int_{|r_1 - r_2|}^{r_1 + r_2} dr \; e^{-\zeta r}$$

$$= \frac{1}{2\zeta r_1 r_2} \left[e^{-\zeta |r_1 - r_2|} - e^{-\zeta (r_1 + r_2)} \right] \tag{F.18}$$

which can be compared with equation (F.14). For the screened Coulomb potential, where the potential F is given by

$$F(|\mathbf{x}_1 - \mathbf{x}_2|) = \frac{e^{-\zeta |\mathbf{x}_1 - \mathbf{x}_2|}}{|\mathbf{x}_1 - \mathbf{x}_2|}$$

$$= \sum_l f_l(r_1, r_2) P_l(\hat{\mathbf{x}}_1 \cdot \hat{\mathbf{x}}_2) \tag{F.19}$$

we can notice that when $|\mathbf{x}_1 - \mathbf{x}_2| \neq 0$, F satisfies the Helmholtz equation:

$$\left[-\Delta + \zeta^2 \right] F(|\mathbf{x}_1 - \mathbf{x}_2|) = \sum_{l=0}^{\infty} \left[-\Delta + \zeta^2 \right] f_l(r_1, r_2) P_l(\hat{\mathbf{x}}_1 \cdot \hat{\mathbf{x}}_2) = 0 \tag{F.20}$$

Since each term in the sum must vanish separately, we have

$$\left[-\Delta + \zeta^2 \right] f_l(r_1, r_2) P_l(\hat{\mathbf{x}}_1 \cdot \hat{\mathbf{x}}_2) = 0 \tag{F.21}$$

from which it follows that

$$\left[r_1^2 \frac{\partial^2}{\partial r_1^2} + 2r_1 \frac{\partial}{\partial r_1} - l(l+1) - (\zeta r_1)^2 \right] f_l(r_1, r_2) = 0 \tag{F.22}$$

Similarly

$$\left[r_2^2 \frac{\partial^2}{\partial r_2^2} + 2r_2 \frac{\partial}{\partial r_2} - l(l+1) - (\zeta r_2)^2 \right] f_l(r_1, r_2) = 0 \tag{F.23}$$

Thus $f_l(r_1, r_2)$ obeys the modified spherical Bessel differential equation both with respect to r_1 and with respect to r_2. One can show in general that

$$f_l(r_1, r_2) = (2l + 1)\zeta \; i_l(\zeta r_<) k_l(\zeta r_>) \tag{F.24}$$

where $i_l(x)$ and $k_l(x)$ are modified spherical Bessel functions of the first and second kinds. Modified spherical Bessel functions of the first kind are related to the usual spherical Bessel functions by

$$i_l(x) = i^{-l} j_l(ix) \tag{F.25}$$

while modified spherical Bessel functions of the second kind are given by

$$k_0(x) = \frac{e^{-x}}{x}$$

$$k_1(x) = \frac{e^{-x}(x+1)}{x^2}$$

$$k_2(x) = \frac{e^{-x}(x^2+3x+3)}{x^3}$$

$$\vdots \quad \vdots \quad \vdots \tag{F.26}$$

with the recursion relation

$$k_{l+1}(x) = \frac{2l+1}{x} k_l(x) + k_{l-1}(x) \tag{F.27}$$

The reader may verify that the special cases $f_0(r_1, r_2)$ shown in equations (F.18) has the general form shown in equation (F.24).

F.4 Expansion of a displaced Slater-type orbital

It is interesting to notice that

$$-\frac{\partial}{\partial \zeta} \left[\frac{e^{-\zeta|\mathbf{x}_1 - \mathbf{x}_2|}}{|\mathbf{x}_1 - \mathbf{x}_2|} \right] = -\sum_l P_l(\hat{\mathbf{x}}_1 \cdot \hat{\mathbf{x}}_2) \frac{\partial}{\partial \zeta} [f_l(r_1, r_2)] \tag{F.28}$$

so that

$$e^{-\zeta|\mathbf{x}_1 - \mathbf{x}_2|} = -\sum_l P_l(\hat{\mathbf{x}}_1 \cdot \hat{\mathbf{x}}_2) \frac{\partial}{\partial \zeta} [f_l(r_1, r_2)] \tag{F.29}$$

where $f_l(r_1, r_2)$ is given by equation (F.24). Thus we can write

$$e^{-\zeta|\mathbf{x}_1 - \mathbf{x}_2|} = \sum_l \hat{f}_l(r_1, r_2) P_l(\hat{\mathbf{x}}_1 \cdot \hat{\mathbf{x}}_2) \tag{F.30}$$

where

$$\hat{f}_l(r_1, r_2) = -(2l+1) \frac{\partial}{\partial \zeta} [\zeta \, i_l(\zeta r_<) k_l(\zeta r_>)] \tag{F.31}$$

The differentiation may be performed by means of the relationships

$$(2l+1) i_l'(x) = l i_{l-1}(x) + (l+1) i_{l+1}(x) \tag{F.32}$$

and

$$(2l+1) k_l'(x) = -l k_{l-1}(x) - (l+1) k_{l+1}(x) \tag{F.33}$$

For other STO's, expansion of the displaced orbitals can be obtained by successive differentiation with respect to ζ.

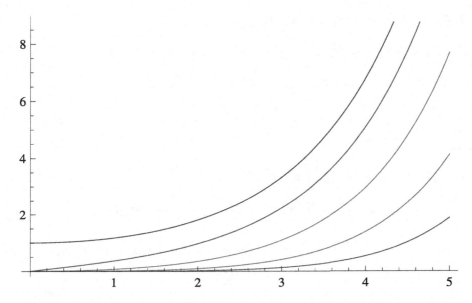

Fig. F.1 *Modified spherical Bessel functions of the first kind, $i_0(x)$ (uppermost curve),*
$i_1(x)$, $i_2(x)$, $i_3(x)$ and $i_4(x)$ (lowest curve).

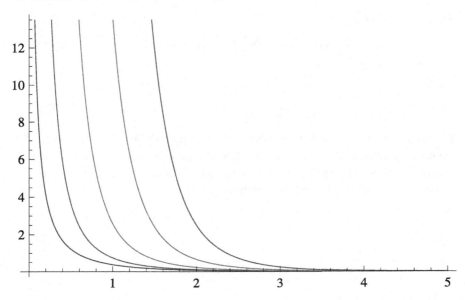

Fig. F.2 *Modified spherical Bessel functions of the second kind, $k_0(x)$ (lefthand curve),*
$k_1(x)$, $k_2(x)$, $k_3(x)$ and $k_4(x)$ (farthest to the right).

F.5   A Fourier transform solution

Let $F(|\mathbf{x}_1 - \mathbf{x}_2|)$ be any function of the distance $|\mathbf{x}_1 - \mathbf{x}_2|$, and let

$$
\begin{aligned}
F^t(p) &= \frac{1}{(2\pi)^{3/2}} \int d^3x \ e^{-i\mathbf{p}\cdot\mathbf{x}} F(r) \\
&= \sqrt{\frac{2}{\pi}} \int_0^\infty dr \ r^2 j_0(pr) F(r)
\end{aligned}
\tag{F.34}
$$

Then

$$
\begin{aligned}
F(|\mathbf{x}_1 - \mathbf{x}_2|) &= \frac{1}{(2\pi)^{3/2}} \int d^3p \ e^{i\mathbf{p}\cdot(\mathbf{x}_1-\mathbf{x}_2)} F^t(p) \\
&= \frac{1}{(2\pi)^{3/2}} \sum_l i^l (2l+1) \int_0^\infty dp \ p^2 j_l(pr_1) F^t(p) \\
&\quad \times \sum_{l'} (-i)^{l'} (2l'+1) j_{l'}(pr_2) \int d\Omega_p \ P_{l'}(\hat{\mathbf{p}}\cdot\hat{\mathbf{x}}_2) P_l(\hat{\mathbf{p}}\cdot\hat{\mathbf{x}}_2) \\
&= \sum_l P_l(\hat{\mathbf{x}}_1\cdot\hat{\mathbf{x}}_2)(2l+1)\sqrt{\frac{2}{\pi}} \int_0^\infty dp \ p^2 j_l(pr_1) j_l(pr_2) F^t(p)
\end{aligned}
\tag{F.35}
$$

Thus we can write

$$
F(|\mathbf{x}_1 - \mathbf{x}_2|) = \sum_l f_l(r_1, r_2) P_l(\hat{\mathbf{x}}_1 \cdot \hat{\mathbf{x}}_2)
\tag{F.36}
$$

where

$$
f_l(r_1, r_2) = (2l+1)\sqrt{\frac{2}{\pi}} \int_0^\infty dp \ p^2 j_l(pr_1) j_l(pr_2) F^t(p)
\tag{F.37}
$$

The methods discussed here for expanding displaced functions in terms of Legendre polynomials may be applied to the problem of evaluating many-center interelectron repulsion integrals when Coulomb Sturmians are used as basis functions for molecular calculations.

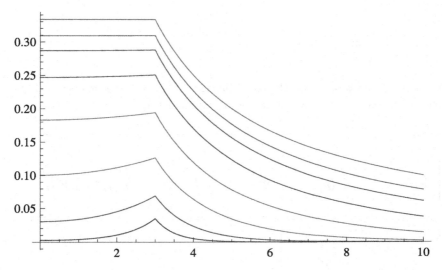

Fig. F.3 *The function $f_l(r_1, r_2) = (2l+1)\zeta\, i_l(\zeta r_<)k_l(\zeta r_<)$ compared with $r_<^l/r_>^{l+1}$ for $l = 0$, $r_2 = 3$ and for various values of the screening constant ζ. As $\zeta \to 0$, $f_0(r_1, r_2)$ approaches $1/r_>$ (top curve). The values of ζ used were .025, .05, .1, .2, .4, .8, and 1.6, with 1.6 corresponding to the bottom curve.*

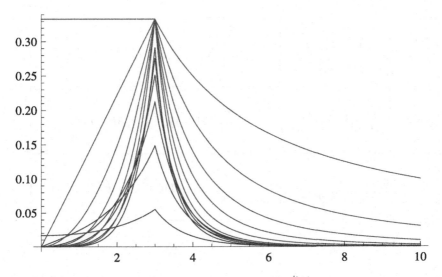

Fig. F.4 *This figure shows $f_l(r_1, r_2)$ compared with $r_<^l/r_>^{l+1}$ for $r_2 = 3$ for $\zeta = 1$ and various values of l. For very large values of l, $f_l(r_1, r_2)$ approaches $r_<^l/r_>^{l+1}$.*

Table F.1 This table shows the radial functions $f_l(r_<, r_>)$ that appear in Legendre polynomial expansion of a displaced function $F(r)$.

$F(r)$	$F^t(p)$	$f_l(r_<, r_>) =$ $(2l+1)\sqrt{\dfrac{2}{\pi}} \displaystyle\int_0^\infty dp\, p^2 j_l(pr_1) j_l(pr_2) F^t(p)$
$\dfrac{1}{r}$	$\sqrt{\dfrac{2}{\pi}} \dfrac{1}{p^2}$	$\dfrac{r_<^l}{r_>^{l+1}}$
$\dfrac{e^{-\zeta r}}{r}$	$\sqrt{\dfrac{2}{\pi}} \dfrac{1}{p^2 + \zeta^2}$	$(2l+1)\zeta\, i_l(\zeta r_<) k_l(\zeta r_>)$
$e^{-\zeta r}$	$-\dfrac{\partial}{\partial \zeta} \sqrt{\dfrac{2}{\pi}} \dfrac{1}{p^2 + \zeta^2}$	$-\dfrac{\partial}{\partial \zeta}[(2l+1)\zeta\, i_l(\zeta r_<) k_l(\zeta r_>)]$
$r e^{-\zeta r}$	$\dfrac{\partial^2}{\partial \zeta^2} \sqrt{\dfrac{2}{\pi}} \dfrac{1}{p^2 + \zeta^2}$	$\dfrac{\partial^2}{\partial \zeta^2}[(2l+1)\zeta\, i_l(\zeta r_<) k_l(\zeta r_>)]$
$r^2 e^{-\zeta r}$	$-\dfrac{\partial^3}{\partial \zeta^3} \sqrt{\dfrac{2}{\pi}} \dfrac{1}{p^2 + \zeta^2}$	$-\dfrac{\partial^3}{\partial \zeta^3}[(2l+1)\zeta\, i_l(\zeta r_<) k_l(\zeta r_>)]$

F.6 Displacement of functions that do not have spherical symmetry

It might be asked whether the method of Legendre polynomial expansions can be used to represent displaced functions that do not have spherical symmetry. For example, consider the function

$$F(r)x_1^{n_1} x_2^{n_2} x_3^{n_3} \tag{F.38}$$

When displaced from a point $\mathbf{B}$ to the origin, this function can be represented as

$$(x_1 - B_1)^{n_1} (x_2 - B_2)^{n_2} (x_3 - B_3)^{n_3} \sum_l f_l(r, B) P_l(\hat{\mathbf{x}} \cdot \hat{\mathbf{B}})$$

$$= \frac{1}{n!} \frac{\partial^n}{\partial A_1^{n_1} \partial A_2^{n_2} \partial A_3^{n_3}} [\mathbf{A} \cdot (\mathbf{x} - \mathbf{B})]^n \sum_l f_l(r, B) P_l(\hat{\mathbf{x}} \cdot \hat{\mathbf{B}}) \tag{F.39}$$

where $n = n_1 + n_2 + n_3$. Angular integrals involving functions of the form

$$[\mathbf{A} \cdot (\mathbf{x} - \mathbf{B})]^n P_l(\hat{\mathbf{x}} \cdot \hat{\mathbf{B}}) \tag{F.40}$$

can easily be evaluated using the vector pairing techniques discussed in Appendix C and in references [Avery and Ørmen, 1980] and [Michels, 1981]. The methods discussed here can be applied to the evaluation of many-center interelectron repulsion integrals for Exponential-Type Orbitals (ETO's).

Appendix G

MULTIPOLE EXPANSIONS

The electrostatic interaction between a charge density ρ_a, centered at the point $\mathbf{X}_a$, and another charge density ρ_b, centered at the point $\mathbf{X}_b$, is given by

$$J = \int d^3x \int d^3x' \; \rho_a(\mathbf{x} - \mathbf{X}_a)\rho_b(\mathbf{x}' - \mathbf{X}_b)\frac{1}{|\mathbf{x} - \mathbf{x}'|} \qquad (G.1)$$

If we introduce the notation

$$\mathbf{x}_a \equiv \mathbf{x} - \mathbf{X}_a$$
$$\mathbf{x}_b \equiv \mathbf{x}' - \mathbf{X}_b$$
$$\mathbf{R} \equiv \mathbf{X}_b - \mathbf{X}_a$$
$$R \equiv |\mathbf{X}_b - \mathbf{X}_a|$$
$$\mathbf{x}_{ab} \equiv \mathbf{x}_a - \mathbf{x}_b$$
$$r_{ab} \equiv |\mathbf{x}_a - \mathbf{x}_b| \qquad (G.2)$$

we can rewrite the interelectron repulsion integral as

$$J = \int d^3x \int d^3x' \; \rho_a(\mathbf{x} - \mathbf{X}_a)\rho_b(\mathbf{x}' - \mathbf{X}_b)\frac{1}{|\mathbf{x} - \mathbf{x}'|}$$
$$= \int d^3x_a \int d^3x_b \; \rho_a(\mathbf{x}_a)\rho_b(\mathbf{x}_b)\frac{1}{|\mathbf{x}_a - \mathbf{x}_b + \mathbf{X}_a - \mathbf{X}_b|}$$
$$= \int d^3x_a \int d^3x_b \; \rho_a(\mathbf{x}_a)\rho_b(\mathbf{x}_b)\frac{1}{|\mathbf{x}_{ab} - \mathbf{R}|}$$
$$= \int d^3x_a \int d^3x_b \; \rho_a(\mathbf{x}_a)\rho_b(\mathbf{x}_b)\sum_{l=0}^{\infty} \frac{r_<^l}{r_>^{l+1}} P_l(\hat{\mathbf{R}} \cdot \hat{\mathbf{x}}_{ab}) \qquad (G.3)$$

where

$$\frac{r_<^l}{r_>^{l+1}} \equiv \begin{cases} r_{ab}^l/R^{l+1} & r_{ab} < R \\[2mm] R^l/r_{ab}^{l+1} & r_{ab} > R \end{cases} \qquad (G.4)$$

In the asymptotic region, where the charge distributions do not overlap appreciably, we can assume that $r_{ab} < R$, and therefore we can write

$$J \to \sum_{l=0}^{\infty} \frac{1}{R^{l+1}} \int d^3x_a \int d^3x_b \; \rho_a(\mathbf{x}_a)\rho_b(\mathbf{x}_b) r_{ab}^l P_l(\hat{\mathbf{R}} \cdot \hat{\mathbf{x}}_{ab})$$

$$= \sum_{l=0}^{\infty} \frac{1}{R^{l+1}} \int d^3x_a \int d^3x_b \; \rho_a(\mathbf{x}_a)\rho_b(\mathbf{x}_b) r_{ab}^l \sum_{m=-l}^{l} \frac{4\pi}{2l+1} Y_{l,m}^*(\hat{\mathbf{R}}) Y_{l,m}(\hat{\mathbf{x}}_{ab})$$

$$= \sum_{l=0}^{\infty} \sum_{m=-l}^{l} \frac{1}{R^{l+1}} \frac{4\pi}{2l+1} Y_{l,m}^*(\hat{\mathbf{R}}) \int d^3x_a \int d^3x_b \; \rho_a(\mathbf{x}_a)\rho_b(\mathbf{x}_b) r_{ab}^l Y_{l,m}(\hat{\mathbf{x}}_{ab})$$

$$(G.5)$$

We now introduce the regular and irregular solid harmonics R_l^m and I_l^m, defined by

$$\mathsf{R}_l^m(\mathbf{x}) \equiv \sqrt{\frac{4\pi}{2l+1}} r^l Y_{lm}(\hat{\mathbf{x}})$$

$$\mathsf{I}_l^m(\mathbf{x}) \equiv \sqrt{\frac{4\pi}{2l+1}} \frac{1}{r^{l+1}} Y_{lm}(\hat{\mathbf{x}}) \qquad (G.6)$$

In terms of these, the asymptotic expression for J becomes

$$J \to \sum_{l=0}^{\infty} \sum_{m=-l}^{l} (-1)^m \mathsf{I}_l^{-m}(\mathbf{R}) \int d^3x_a \int d^3x_b \; \rho_a(\mathbf{x}_a)\rho_b(\mathbf{x}_b) \mathsf{R}_l^m(\mathbf{x}_{ab})$$

$$(G.7)$$

The double integral $\int d^3x_a \int d^3x_b$ can be separated into a product of single integrals by means of a standard expansion that makes use of binomial coefficients and Clebsch-Gordan coefficients:

$$\mathsf{R}_l^m(\mathbf{x}_a - \mathbf{x}_b)$$

$$= \sum_{l_a=0}^{l} (-1)^{l-l_a} \binom{2l}{2l_a}^{1/2} \sum_{m_a=-l_a}^{l_a} \mathsf{R}_{l_a}^{m_a}(\mathbf{x}_a) \mathsf{R}_{l-l_a}^{m-m_a}(\mathbf{x}_b) \begin{pmatrix} l_a & l-l_a & l \\ m_a & m-m_a & m \end{pmatrix}$$

$$(G.8)$$

Finally, making the substitutions $\rho_a(\mathbf{x}_a) = \rho_{\mu_1,\mu_2}(\mathbf{x}_a)$ and $\rho_a(\mathbf{x}_b) = \rho_{\mu_3,\mu_4}(\mathbf{x}_b)$, we obtain the asymptotic expression

$$J_{\mu_1,\mu_2;\mu_3,\mu_4}$$

$$\to \sum_{l=0}^{\infty} \sum_{m=-l}^{l} (-1)^m \mathsf{I}_l^{-m}(\mathbf{R}) \sum_{l_a=0}^{l} (-1)^{l-l_a} \binom{2l}{2l_a}^{1/2} \begin{pmatrix} l_a & l-l_a & l \\ m_a & m-m_a & m \end{pmatrix}$$

$$\times \int d^3x_a \; \rho_{\mu_1,\mu_2}(\mathbf{x}_a) \mathsf{R}_{l_a}^{m_a}(\mathbf{x}_a) \int d^3x_b \; \rho_{\mu_3,\mu_4}(\mathbf{x}_b) \mathsf{R}_{l-l_a}^{m-m_a}(\mathbf{x}_b) \qquad (G.9)$$

In the special case where the densities $\rho_a(\mathbf{x}_a) = \rho_{\mu_1,\mu_2}(\mathbf{x}_a)$ and $\rho_a(\mathbf{x}_b) = \rho_{\mu_3,\mu_4}(\mathbf{x}_b)$ are formed by products of Coulomb Sturmian basis functions, this can be rewritten in the form:

$$J_{\mu_1,\mu_2;\mu_3,\mu_4}$$

$$\rightarrow k \sum_{l=0}^{\infty} \sum_{m=-l}^{l} (-1)^m l_l^{-m}(\mathbf{S}) \sum_{l_a=0}^{l} (-1)^{l-l_a} \begin{pmatrix} 2l \\ 2l_a \end{pmatrix}^{1/2} \begin{pmatrix} l_a & l-l_a & l \\ m_a & m-m_a & m \end{pmatrix}$$

$$\times \int d^3 s_a \; \rho_{\mu 1,\mu 2}(\mathbf{s}_a) R_{l_a}^{m_a}(\mathbf{s}_a) \int d^3 s_b \; \rho_{\mu 3,\mu 4}(\mathbf{x}_b) R_{l-l_a}^{m-m_a}(\mathbf{s}_b) \qquad (G.10)$$

where $\mathbf{s}_a \equiv k\mathbf{x}_a$, $\mathbf{s}_b \equiv k\mathbf{x}_b$ and $\mathbf{S} \equiv k\mathbf{R}$.

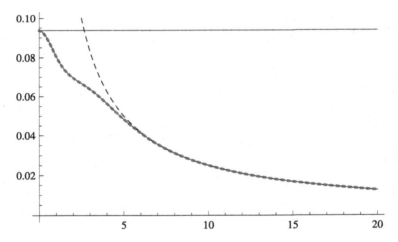

Fig. G.1 *In this figure the asymptotic curve $J_{1s,2s,2s,1s}$ is shown as a function of $S = kR$ for a diatomic molecule, where R is the internuclear distance. The asymptotic curve is shown as a thin dashed line. For comparison, the atomic value of $J_{1s,2s,2s,1s}$ is shown as a thin solid line, while the integral calculated using the hyperspherical method is shown as a line with thick dashes.*

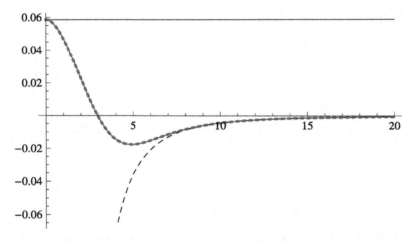

Fig. G.2   *This figure is the same as Figure G.1 except that $J_{2p_0,2s,2p_0,2s}$ is shown.*

Appendix H

HARMONIC FUNCTIONS

H.1 Harmonic functions for $d = 3$

Harmonic functions in a 3-dimensional space are solutions to the Laplace equation

$$\nabla^2 \phi(\mathbf{x}) = 0 \tag{H.1}$$

Let us begin by listing some of these solutions. The regular solid harmonics

$$R_{l,m}(\mathbf{x}) \equiv \sqrt{\frac{4\pi}{2l+1}}\ r^l Y_{l,m}(\hat{\mathbf{x}}) \tag{H.2}$$

satisfy the Laplace equation everywhere in space, while the irregular solid harmonics

$$I_{l,m}(\mathbf{x}) \equiv \sqrt{\frac{4\pi}{2l+1}}\ r^{-(l+1)} Y_{l,m}(\hat{\mathbf{x}}) \tag{H.3}$$

satisfy it everywhere except at the point $r = |\mathbf{x}| = 0$. Here $Y_{l,m}(\hat{\mathbf{x}})$ is a spherical harmonic. From the sum-rule for spherical harmonics,

$$\sum_{m=-l}^{l} Y_{l,m}^*(\hat{\mathbf{a}})Y_{l,m}(\hat{\mathbf{x}}) = \frac{2l+1}{4\pi}P_l(\hat{\mathbf{a}}\cdot\hat{\mathbf{x}}) \tag{H.4}$$

we can see that for any constant vector $\mathbf{a}$, the function

$$r^l P_l'(\hat{\mathbf{a}}\cdot\hat{\mathbf{x}}) \equiv r^l \sum_{m=-l}^{l} \frac{4\pi}{2l+1}Y_{l,m}^*(\hat{\mathbf{a}})Y_{l,m}(\hat{\mathbf{x}})$$

$$= \sum_{t=0}^{[n/2]} \frac{(-1)^t \Gamma(1+1/2-t)(x^2+y^2+z^2)^t(2\hat{\mathbf{a}}\cdot\mathbf{x})^{l-2t}}{t!(1-2t)!\Gamma(1/2)}$$

$$\tag{H.5}$$

is harmonic, i.e. it is a solution to the Laplace equation. The function $P'_l(\hat{\mathbf{a}} \cdot \hat{\mathbf{x}})$ might be called a "harmonic Legendre polynomial". Harmonic polynomials are, by definition, homogeneous polynomials that are solutions to the Laplace equation. The functions $r^l Y_{l,m}(\hat{\mathbf{x}})$ satisfy this definition, and they are thus harmonic polynomials.

The spherical harmonics corresponding to a given value of l form an irreducible representation of SO(3). The point groups of chemistry are subgroups of SO(3), and therefore the spherical harmonics corresponding to a given value of l are closed under the operations of the point groups. Thus if $\mathcal{G}$ is such a point group, each $(2l + 1)$-dimensional set of spherical harmonics, $Y_{l,m}(\hat{\mathbf{x}})$ can be thought of as invariant subset with respect to $\mathcal{G}$ in the sense discussed elsewhere in this book. Similar considerations hold for the hyperspherical harmonics which will be discussed below.

An alternative set of harmonic polynomials, spanning the same part of Hilbert space as the solid harmonics, can be generated by starting with monomials of the form

$$f_{\mathbf{n}}(\mathbf{x}) = x^{n_1} y^{n_2} z^{n_3} \qquad n_j = 0, 1, 2, 3, \ldots \qquad \text{(H.6)}$$

and acting on them several times with the Laplacian operator with appropriate coefficients, as is discussed in Appendix C. We then obtain harmonic polynomials of the form:

$$h_{\mathbf{n}}(\mathbf{x}) = \sum_{j=0}^{\lfloor n/2 \rfloor} \frac{(-1)^j (2n - 2j - 1)!!}{(2j)!!(2n - 1)!!} r^{2j} (\nabla^2)^j f_{\mathbf{n}}(\mathbf{x})$$

$$\text{(H.7)}$$

where n is the degree of the monomial, and also the degree of the harmonic polynomial:

$$n \equiv n_1 + n_2 + n_3 \qquad \text{(H.8)}$$

Finally it should be mentioned that the functions

$$g_{\mathbf{n}}(\mathbf{x}) = r^{-(2n+1)} h_{\mathbf{n}}(\mathbf{x}) \qquad \text{(H.9)}$$

also satisfy the Laplace equation except at the point $r = |\mathbf{x}| = 0$. For some purposes the harmonic functions $h_{\mathbf{n}}(\mathbf{x})$ and $g_{\mathbf{n}}(\mathbf{x})$ may be more convenient than $a_{l,m}(\mathbf{x})$ and $b_{l,m}(\mathbf{x})$.

H.2   Spaces of higher dimension

In a d-dimensional space, the generalized Laplace equation has the form

$$\Delta \phi(\mathbf{x}) = 0 \qquad \text{(H.10)}$$

where

$$\Delta \equiv \sum_{j=1}^{d} \frac{\partial^2}{\partial x_j} \tag{H.11}$$

and where

$$\mathbf{x} \equiv (x_1, x_2, x_3, \ldots, x_d) \tag{H.12}$$

is a vector whose components are the Cartesian coordinates for the space. We can introduce the hyperradius r defined by

$$r^2 \equiv \sum_{j=1}^{d} x_j^2 \tag{H.13}$$

and a generalized angular momentum operator

$$\Lambda^2 \equiv -\sum_{i>j}^{d} \sum_{j=1}^{d} \left(x_i \frac{\partial}{\partial x_j} - x_j \frac{\partial}{\partial x_i} \right)^2 \tag{H.14}$$

Written in terms of the hyperradius and the generalized angular momentum operator, the generalized Laplace operator takes the form

$$\Delta = \frac{1}{r^{d-1}} \frac{\partial}{\partial r} r^{d-1} \frac{\partial}{\partial r} - \frac{\Lambda^2}{r^2} \tag{H.15}$$

Hyperspherical harmonics $Y_{\lambda,\mu}(\hat{\mathbf{x}})$, are the d-dimensional analogues of spherical harmonics. They are defined as eigenfunctions of the generalized angular momentum operator such that

$$\Lambda^2 Y_{\lambda,\mu}(\hat{\mathbf{x}}) = \lambda(\lambda + d - 2) Y_{\lambda,\mu}(\hat{\mathbf{x}}) \tag{H.16}$$

and such that $r^\lambda Y_{\lambda,\mu}(\hat{\mathbf{x}})$ is a homogeneous polynomial. Then

$$\Delta r^\lambda Y_{\lambda,\mu}(\hat{\mathbf{x}}) = \left(\frac{1}{r^{d-1}} \frac{\partial}{\partial r} r^{d-1} \frac{\partial}{\partial r} - \frac{\Lambda^2}{r^2} \right) r^\lambda Y_{\lambda,\mu}(\hat{\mathbf{x}})$$

$$= \left(\frac{1}{r^{d-1}} \frac{\partial}{\partial r} r^{d-1} \frac{\partial r^\lambda}{\partial r} - \frac{\lambda(\lambda + d - 2)}{r^2} r^\lambda \right) Y_{\lambda,\mu}(\hat{\mathbf{x}}) = 0 \tag{H.17}$$

Thus it can be seen that hyperspherical harmonics are defined in such a way that $r^\lambda Y_{\lambda,\mu}(\hat{\mathbf{x}})$ is a harmonic polynomial in the d-dimensional space. The index μ is actually a set of d-2 indices. For hyperspherical harmonics of the standard type, these indices are organized by means of a chain of subgroups:

$$SO(d) \supset SO(d-1) \supset SO(d-2) \supset \cdots \supset SO(2) \tag{H.18}$$

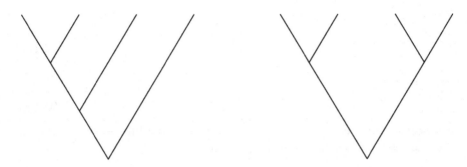

Fig. H.1   The standard tree (left) and an alternative tree (right) for 4-dimensional hyperspherical harmonics. The standard tree on the left corresponds to the ordering of subgroups shown in equation (H.18), in which a harmonic polynomials are first found in the space spanned by the coordinates x_1 and x_2. These are then multiplied by x_3 to form homogeneous polynomials of degree 3, from which the harmonic parts are projected out. Finally these are coupled to x_4. The right-hand tree symbolizes the alternative scheme of equation (H.19), where harmonic polynomials are first constructed within the subspaces (x_1, x_2) and (x_3, x_4). Finally these are coupled together, and polynomials that are harmonic in the entire 4-dimensional space are obtained. 4-dimensional hyperspherical harmonics corresponding to the standard tree are shown in Table 5.1, while those corresponding to the alternative tree are shown in Tables H.1 and H.2.

However, when we use hyperspherical harmonics in physical problems, it may be convenient to organize the minor indices μ according to a different chain of subgroups. Tables H.1 and H.2 show an alternative set of 4-dimensional hyperspherical harmonics where the minor indices are organized according to the chain

$$SO(4) \supset SO(2) \times SO(2) \tag{H.19}$$

This alternative way of organizing the minor indices is symbolized by the right-hand tree in Figure H.1. An excellent discussion of the method of trees in hyperspherical harmonic theory can be found in [Coletti, 1998].

The hyperspherical harmonics corresponding to a given value of the principal quantum number λ form an invariant subspace with respect to all groups $\mathcal{G}$ that are subgroups of $SO(d)$. In the case where $d = 3$, there are $2l + 1$ linearly independent functions in this subspace. In the general case, the corresponding number of linearly independent functions in the invariant subspace can be shown to be [Avery, 1989]

$$m_\lambda = \frac{(d + 2\lambda - 2)(d + \lambda - 3)!}{\lambda!(d - 2)!} \tag{H.20}$$

The reader can verify that for $d = 3$ and $\lambda = l$, this reduces to the familiar result that there are $2l + 1$ linearly independent functions. When $d = 4$, the dimension of the invariant subspace is $(\lambda + 1)^2$. Thus for $\lambda = 0, 1, 2, 3, \ldots$ there are respectively $1, 4, 9, 16, \ldots$ *linearly* independent solutions, as is illustrated in Tables H.1, H.2 and 5.1. These dimensions correspond, through Fock's projection (Appendix B), to the number of degenerate hydrogenlike orbitals with principal quantum numbers $n = 1, 2, 3, 4, \ldots$. Thus Fock's projection casts light on the puzzling n^2-fold degeneracy of the hydrogenlike orbitals.

Functions of the form

$$\mathcal{H}_{\lambda,\mu}(\mathbf{x}) = r^\lambda Y_{\lambda,\mu}(\hat{\mathbf{x}}) \tag{H.21}$$

satisfy the generalized Laplace equation everywhere in our d-dimensional space, while functions of the form

$$\mathcal{I}_{\lambda,\mu}(\mathbf{x}) = r^{-\lambda-d+2} Y_{\lambda,\mu}(\hat{\mathbf{x}}) \tag{H.22}$$

satisfy it at all points except the origin. Equations (H.6)-(H.9) also have d-dimensional analogues:

$$f_{\mathbf{n}}(\mathbf{x}) = \prod_{j=1}^{d} x_j^{n_j} \qquad n_j = 0, 1, 2, 3, \ldots \tag{H.23}$$

$$h_{\mathbf{n}}(\mathbf{x}) = \sum_{j=0}^{\lfloor n/2 \rfloor} \frac{(-1)^j (d + 2n - 2j - 4)!!}{(2j)!!(d + 2n - 4)!!} r^{2j} \Delta^j f_{\mathbf{n}}(\mathbf{x}) \tag{H.24}$$

$$n = \sum_{j=1}^{d} n_j \tag{H.25}$$

and

$$g_{\mathbf{n}}(\mathbf{x}) = r^{-2n-d+2} h_{\mathbf{n}}(\mathbf{x}) \tag{H.26}$$

The function $h_{\mathbf{n}}(\mathbf{x})$ satisfies the generalized Laplace equation everywhere in space, while $g_{\mathbf{n}}(\mathbf{x})$ is a solution everywhere except at the origin.

The harmonic functions discussed above by no means exhaust the forms that solutions to the generalized Laplace equation in a d-dimensional space can take. Examples of other forms include

$$e^{i\mathbf{k}\cdot\mathbf{x}} \tag{H.27}$$

where $\mathbf{k}$ is a d-dimensional vector of zero length. We can see that this will be a harmonic function because

$$\Delta e^{i\mathbf{k}\cdot\mathbf{x}} = -\mathbf{k}\cdot\mathbf{k}\, e^{i\mathbf{k}\cdot\mathbf{x}} = 0 \tag{H.28}$$

As an example of a d-dimensional vector of zero length we can think of

$$\mathbf{k} = (k_1, k_2, k_3, \ldots, k_{d-1}, \pm i k_d)$$
$$k_d = \left(k_1^2 + k_2^2 + \cdots + k_{d-1}^2\right)^{1/2}$$
$$\mathbf{k} \cdot \mathbf{k} = k_d^2 - k_d^2 = 0 \qquad (\mathrm{H.29})$$

In fact, if $\mathbf{k}$ is a d-dimensional vector of zero length, any well-behaved function of $\zeta \equiv \mathbf{k} \cdot \mathbf{x}$ will be a solution to the generalized Laplace equation, because

$$\frac{\partial}{\partial x_j} F(\mathbf{k} \cdot \mathbf{x}) = \frac{\partial}{\partial x_j} F(\zeta) = \frac{\partial \zeta}{\partial x_j} \frac{dF}{d\zeta} = k_j \frac{dF}{d\zeta} \qquad (\mathrm{H.30})$$

and

$$\frac{\partial^2}{\partial x_j^2} F(\mathbf{k} \cdot \mathbf{x}) = k_j \frac{\partial}{\partial x_j} \frac{dF}{d\zeta} = k_j \frac{\partial \zeta}{\partial x_j} \frac{d^2 F}{d\zeta^2} = k_j^2 \frac{d^2 F}{d\zeta^2} \qquad (\mathrm{H.31})$$

Thus

$$\Delta F(\mathbf{k} \cdot \mathbf{x}) = \mathbf{k} \cdot \mathbf{k} \, \frac{d^2 F}{d\zeta^2} = 0 \qquad (\mathrm{H.32})$$

In a d-dimensional space, it is possible to define a "harmonic Gegenbauer polynomial"

$$r^\lambda C_\lambda'^\alpha(\hat{\mathbf{a}} \cdot \hat{\mathbf{x}}) \equiv \sum_{t=0}^{[\lambda/2]} \frac{(-1)^t \Gamma(\lambda + \alpha - t)(x_1^2 + x_2^2 + \cdots + x_d^2)^t (2\hat{\mathbf{a}} \cdot \mathbf{x})^{l-2t}}{t!(\lambda - 2t)!\Gamma(\alpha)}$$
$$(\mathrm{H.33})$$

where $\alpha = d/2 - 1$. The harmonic Gegenbauer polynomial is the d-dimensional generalization of the harmonic Legendre polynomial, and it satisfies the generalized Laplace equation for any constant d-dimensional vector $\mathbf{a}$.

Table H.1 Alternative 4-dimensional hyperspherical harmonics corresponding to the right-hand tree in Figure H.1. The indices m_1 and m_2 are rotational quantum numbers in the subspaces spanned respectively by (x_1, x_2) and (x_3, x_4).

λ	m_1	m_2	$\sqrt{2}\pi\, Y_{\lambda,m_1,m_2}(\mathbf{u})$
0	0	0	1
1	1	0	$\sqrt{2}(u_1 + iu_2)$
1	-1	0	$\sqrt{2}(u_1 - iu_2)$
1	0	1	$\sqrt{2}(u_3 + iu_4)$
1	0	-1	$\sqrt{2}(u_3 - iu_4)$

Table H.2 Alternative 4-dimensional hyperspherical harmonics (continued).

λ	m_1	m_2	$\sqrt{2}\pi\, Y_{\lambda,m_1,m_2}(\mathbf{u})$
2	2	0	$\sqrt{3}(u_1 + iu_2)^2$
2	-2	0	$\sqrt{3}(u_1 - iu_2)^2$
2	0	2	$\sqrt{3}(u_3 + iu_4)^2$
2	0	-2	$\sqrt{3}(u_3 - iu_4)^2$
2	1	1	$\sqrt{6}(u_1 + iu_2)(u_3 + iu_4)$
2	1	-1	$\sqrt{6}(u_1 + iu_2)(u_3 - iu_4)$
2	-1	1	$\sqrt{6}(u_1 - iu_2)(u_3 + iu_4)$
2	-1	-1	$\sqrt{6}(u_1 - iu_2)(u_3 - iu_4)$
2	0	0	$\sqrt{3}(u_1^2 + u_2^2 - u_3^2 - u_4^2)$

Bibliography

Abragam, A. and Bleany B., **Electron Paramagnetic Resonance of Transition Ions**, Clarendon, 1970.

Akhiezer, A.I. and Berestetskii, V.B., **Quantum Electrodynamics**, Interscience, New York, 1965.

Anderson, R.W.; Aquilanti, V.; Cavalli, S. and Grossi, G., *Stereodirected states in molecular dynamics: a discrete basis for the quantum mechanical scattering matrix*, J. Phys. Chem., **95** 8184, 1991.

Aquilanti, V., Cavalli, S., De Fazio, D, and Grossi, G. *Hyperangular momentum: Applications to atomic and molecular science*, in **New Methods in Quantum Theory**, Tsipis, C.A., Popov, V.S., Herschbach, D.R., and Avery, J.S., Eds., Kluwer, Dordrecht, 1996.

Aquilanti, V.; Cavalli, S. and Grossi, G., *Hyperspherical coordinates for molecular dynamics by the method of trees and the mapping of potential-energy surfaces for triatomic systems*, J. Chem. Phys., **85** 1362, 1986.

Aquilanti, V.; Grossi, G.; Laganá, A.; Pelikan, E., and Klar, H., *A decoupling scheme for a 3-body problem treated by expansions into hyperspherical harmonics. The hydrogen molecular ion*, Lett. Nuovo Cimento, **41** 541, 1984.

Aquilanti, V.; Grossi, G., and Laganá, A., *On hyperspherical mapping and harmonic expansions for potential energy surfaces* J. Chem. Phys., **76** 1587-8, 1982.

Aquilanti, V. and Cavalli, S., *Hyperspherical analysis of kinetic paths for elementary chemical reactions and their angular momentum dependence*, Chem. Phys. Lett., **141** 309, 1987.

Aquilanti, V.; Cavalli, S.; Grossi, G., and Anderson, R.W., *Representation in Hyperspherical and Related Coordinates of the potential-energy surface for triatomic reactions*, J. Chem. Soc. Faraday Trans., **86** 1681, 1990.

Aquilanti, V.; Benevente, L.; Grossi, G. and Vecchiocattivi, F., *Coupling schemes for atom-diatom interactions, and an adiabatic decoupling treatment of rotational temperature effects on glory scattering*, J. Chem. Phys., **89** 751-761, 1988.

Aquilanti, V. and Grossi, G., *Angular momentum coupling schemes in the quantum mechanical treatment of P-state atom collisions* J. Chem. Phys., **73**

1165-1172, 1980.

Aquilanti, V. and Grossi, G., *Discrete representations by artificial quantization in the quantum mechanics of anisotropic interactions*, Lett. Nuovo Cimento, **42** 157, 1985.

Aquilanti, V.; Cavalli, S., and Grossi, G., *Discrete analogs of spherical harmonics and their use in quantum mechanics: The hyperquantization algorithm*, Theor. Chem. Acta, **79** 283, 1991.

Aquilanti, V. and Cavalli, S., *Discrete Analogs of Hyperspherical Harmonics and their use for the Quantum Mechanical Three Body Problem*, Few Body Systems, Suppl., **6** 573, 1992.

Aquilanti, V. Cavalli, S. and De Fazio, D., *Angular and hyperangular momentum coupling coefficients as Hahn polynomials*, J. Phys. Chem., **99** 15694, 1995.

Aquilanti, V., Cavalli, S., Coletti, C. and Grossi, G., *Alternative Sturmian bases and momentum space orbitals; an application to the hydrogen molecular ion*, Chem. Phys., **209** 405, 1996.

Aquilanti, V., Cavalli, S. and Coletti, C., *The d-dimensional hydrogen atom; hyperspherical harmonics as momentum space orbitals and alternative Sturmian basis sets*, Chem. Phys., **214** 1, 1997.

Aquilanti, V., and Avery, J., *Generalized potential harmonics and contracted sturmians*, Chem. Phys. Letters, **267** 1, 1997.

Aquilanti, V., Cavalli, S., Coletti, C., Di Domenico, D., and Grossi, G., *Hyperspherical harmonics as Sturmian orbitals in momentum space: a systematic approach to the few-body Coulomb problem*, Int. Rev. Phys. Chem., **20** 673, 2001.

Aquilanti, V. and Caligiana, A., *Sturmian approach to one-electron many-center systems: integrals and iteration schemes*, Chem. Phys. Letters, **366** 157 2002.

Aquilanti, V., Caligiana A., and Cavalli, S., *Hydrogenic elliptic orbitals, coulomb sturmian sets. Recoupling coefficients among alternative bases*, Int. J. Quantum Chem., **92** 99, 2003.

Aquilanti, V., Caligiana, A., Cavalli, S., and Coletti, C., *Hydrogenic orbitals in momentum space and hyperspherical harmonics. Elliptic sturmian basis sets*, Int. J. Quantum Chem., **92** 212, 2003.

Aquilanti, V., and Caligiana, A., *Sturmian orbitals in Quantum Chemistry: an introduction*, in **Fundemental World of Quantum Chemistry: A Tribute to the Memory of P.O. Löwdin**, I, Brändas, E.J. and Kryachko, E.S., Eds., Kluwer, Dordrecht, 297, 2003.

Aquilanti, V. and Avery, J., *Sturmian expansions for quantum mechanical many-body problems and hyperspherical harmonics*, Adv. Quant. Chem., **39** 72-101, 2001.

Atkins, P.W., Child, M.S. and Phillips, C.S.G., **Tables for Group Theory**, Oxford, 1970.

Avery, J., **Creation and Annihilation Operators**, McGraw-Hill, 1976.

Avery, J. and Ørmen, P.J., *Transferable integrals in a deformation density approach to molecular orbital calculations; IV. Evaluation of angular integrals by a vector-pairing method*, Int. J. Quantum Chem., **18** 953-971, 1980.

Avery, J., **Hyperspherical Harmonics; Applications in Quantum Theory**, Kluwer Academic Publishers, Dordrecht, 1989.

Avery, J., *Hyperspherical Sturmian basis functions in reciprocal space*, in **New Methods in Quantum Theory**, Tsipis, C.A., Popov, V.S., Herschbach, D.R., and Avery, J.S., Eds., Kluwer, Dordrecht, 1996.

Avery, J. and Antonsen, F., *A new approach to the quantum mechanics of atoms and small molecules*, Int. J. Quantum Chem., Symp., **23** 159, 1989.

Avery, J. and Antonsen, F., *Iteration of the Schrödinger equation, starting with Hartree-Fock wave functions*, Int. J. Quantum Chem., **42** 87, 1992.

Avery, J. and Antonsen, F., Theor. Chim. Acta, **85** 33, 1993.

Avery, J. and Herschbach, D. R., *Hyperspherical Sturmian basis functions*, Int. J. Quantum Chem., **41** 673, 1992.

Avery, J. and Wen, Z.-Y., *A Formulation of the quantum mechanical many-body in terms of hyperspherical coordinates*, Int. J. Quantum Chem., **25** 1069, 1984.

Avery, J., *Correlation in iterated solutions of the momentum-space Schrödinger equation*, Chem. Phys. Lett., **138** (6) 520-4, 1987.

Avery, J., *Hyperspherical harmonics; Some properties and applications*, in **Conceptual Trends in Quantum Chemistry**, Kryachko, E.S., and Calais, J.L., Eds, Kluwer, Dordrecht, 1994.

Avery, J., Hansen, T.B., Wang, M. and Antonsen, F., *Sturmian basis sets in momentum space*, Int. J. Quantum Chem., **57** 401, 1996.

Avery, J., and Hansen, T.B., *A momentum-space picture of the chemical bond* Int. J. Quantum Chem., **60** 201, 1996.

Avery, J., *Many-particle Sturmians*, J. Math. Chem., **21** 285, 1997.

Avery, J. and Antonsen, F., *Relativistic sturmian basis functions*, J. Math. Chem., **24** 175, 1998.

Avery, J., *A formula for angular and hyperangular integration*, J. Math. Chem., **24** 169, 1998.

Avery, J., *Many-electron Sturmians applied to atoms and ions*, J. Mol. Struct., **458** 1, 1999.

Avery, J., *Many-electron Sturmians as an alternative to the SCF-CI Method*, Adv. Quantum Chem., **31** 201, 1999.

Avery, J., **Hyperspherical Harmonics and Generalized Sturmians**, Kluwer Academic Publishers, Dordrecht, Netherlandsf, 2000.

Avery, J. *Selected applications of hyperspherical harmonics in quantum theory*, J. Phys. Chem.. *97* 2406, 1993.

Avery, J. and Antonsen, F., *Evaluation of angular integrals by harmonic projection*, Theor. Chim. Acta., **85** 33, 1993.

Avery, J., *Fock transforms in reciprocal-space quantum theory*, J. Math. Chem., **15** 233, 1994.

Aquilanti V. and Avery, J., *Generalized potential harmonics and contracted Sturmians*, Chem. Phys. Letters, **267** 1-8, 1997.

Avery, J., *Many-electron Sturmians applied to atoms and ions*, J. Mol. Struct. (Theochem), **458** 1-9, 1999.

Avery, J., *A formula for angular and hyperangular integration*, J. Math. Chem.,

24 169-174, 1998.

Avery, J. and Sauer, S., *Many-electron Sturmians applied to molecules*, in **Quantum Systems in Chemistry and Physics, Volume 1**, Hernández-Laguna, A., Maruani, J., McWeeney, R. and Wilson, S, Eds., Kluwer Academic Publishers, 2000.

Avery, J. and Coletti, C., *Generalized Sturmians applied to atoms in strong external fields*, J. Math. Chem., **27** 43-51 2000.

Avery, J. and Coletti, C., *Many-electron Sturmians applied to atoms and ions in strong external fields*, in **New Trends in Quantum Systems in Chemistry and Physics**, Marauani, J. et al., Eds., 77-93, Kluwer Academic Publishers, 2001.

Avery, J. and Shim, R., *Core ionization energies of atoms and molecules calculated using the Generalized Sturmian Method*, Int. J. Quantum Chem., **79** 1-7, 2000.

Avery, J., *The Generalized Sturmian Method and inelastic scattering of fast electrons*, J. Math. Chem., **27**, 279-292, 2000.

Avery, J., *Sturmian methods in quantum theory*, in **Proc. Workshop on Concepts in Chemical Physics**, Billing, G.D. and Henriksen, N., Eds., Danish Technical University, 2001.

Avery, J., *Sturmians*, in **Handbook of Molecular Physics and Quantum Chemistry**, Wilson, S., Ed., Wiley, Chichester, 2003.

Avery, J., *Harmonic polynomials, hyperspherical harmonics, and Sturmians*, in **Fundemental World of Quantum Chemistry, A Tribute Volume to the Memory of Per-Olov Löwdin**, Brändas, E.J. and Kryachko, E.S., Eds., Kluwer Academic Publishers, Dordrecht, 261-296, 2003.

Avery, J. and Avery, J., *The Generalized Sturmian Method for calculating spectra of atoms and ions*, J. of Math. Chem., **33** 145-162, 2003.

Avery, J. and Avery, J., *Kramers pairs in configuration interaction*, Adv. Quant. Chem., **43** 185-206, 2003.

Avery, J., *Many-center Coulomb Sturmians and Shibuya-Wulfman integrals*, Int. J. Quantum Chem., **100** 121-130, 2003.

Avery, J., Avery J. and Goscinski, O., *Natural orbitals from generalized Sturmian calculations*, Adv. Quant. Chem., **43** 207-16, 2003.

Avery, J., Avery, J., Aquilanti, V., and Caligiana, A., *Atomic densities, polarizabilities and natural orbitals derived from generalized Sturmian calculations*, Adv. Quant. Chem., **47** 156-173, 2004.

Avery, J. and Avery, J., *Generalized Sturmian solutions for many-particle Schrödinger equations*, J. Phys. Chem. A, **41** 8848, 2004.

Avery, J. and Avery, J., **Generalized Sturmians and Atomic Spectra**, World Scientific, 2006.

Avery, J. and Avery, J., *The Generalized Sturmian Library*, http://sturmian.kvante.org, 2006a.

Avery, J. and Avery, J., *Atomic core-ionization energies; approximately piecewise-linear relationships*, J. Math. Chem., **46** 164-181, 2009.

Avery, J., *Harmonic polynomials, hyperspherical harmonics and atomic spectra*, J. Comp. Appl. Math., 2009.

Avery, J., and Avery, J., *Can Coulomb Sturmians be used as a basis for N-electron molecular calculations?*, J. Phys. Chem A,, **113** 14565, 2009.

Avery, J.E., **New Computational Methods in the Quantum Theory of Nanostructures**, Ph.D. Thesis, University of Copenhagen, 2011.

Balhausen, C.J., **Introduction to Ligand Field Theory**, McGraw-Hill, 1962.

Ballot, L. and Farbre de la Ripelle, M., *Application of the hyperspherical formalism to trinucleon bound-state problems*, Ann. Phys., **127** 62, 1980.

Bandar, M. and Itzyksen, C., *Group theory and the H atom*, Rev. Mod. Phys., **38** 330, 1966.

Bang, J.M. and Vaagen, J.S., *The Sturmian expansion: a well-depth-method for orbitals in a deformed potential*, Z. Phys. A, **297** 223-36, 1980.

Bang, J.M., Gareev, F.G., Pinkston, W.T. and Vaagen, J.S., *One- and two-nucleon overlap functions in nuclear physics* Phys. Rep., **125** 253-399, 1985.

Baryudin, L. E. and Telnov, D. A., *Sturmian expansion of the electron density deformation for 3d-metal ions in electric field*, Vestn. Leningr. Univ., Ser 4: Fiz., Khim., **1** 83-6, 1991.

Baretty, Reinaldo; Ishikawa, Yasuyuki; and Nieves, Jose F., *Momentum space approach to relativistic atomic structure calculations*, Int. J. Quantum Chem., Quantum Chem. Symp., **20** 109-17, 1986.

Biedenharn, L.C. and Louck, J.D., **Angular Momentum in Quantum Physics**, Addison Wesley, Reading, Mass, 1981.

Biedenharn, L.C. and Louck, J.D., **The Racah-Wigner Algebra in Quantum Theory**, Addison Wesley, Reading, Mass, 1981.

Bishop, D.B., **Group Theory and Chemistry**, Courier Dover Publications, 1993.

Blinder, S.M., *On Green's functions, propagators, and Sturmians for the nonrelativistic Coulomb problem*, Int. J. Quantum Chem., Quantum Chem. Symp., **18** 293-307, 1984.

Bransden, B.H.; Noble, C.J.; and Hewitt, R.N., *On the reduction of momentum space scattering equations to Fredholm form*, J. Phys. B: At., Mol., Opt. Phys., **26** 2487-99, 1993.

Brandsen, B.H., and Joachim, C.J., **Physics of Atoms and Molecules; 2nd Ed.**, Prentace-Hall, 2003.

Brink, D.M. and Satchler, D.R., **Angular Momentum**, Clarendon, Oxford, 1968.

Caligiana. A., **Sturmian Orbitals in Quantum Chemistry**, Ph.D. thesis, Department of Chemistry, University of Perugia, Italy, October, 2003.

Coleman, A.J., *The Symmetric Group Made Easy*, Adv. Quantum Chem., **4** 83-108, 1968.

Coletti, C. **Struttura Atomica e Moleculare Come Tottura della Simmetria Ipersferica**, Ph.D. thesis, Department of Chemistry, University of Perugia, Italy, 1998.

Condon, E.U. and Shortley, G.H., **Theory of Atomic Spectra**, Cambridge, 1953.

Cotton, F.A., **Chemical Applications of Group Theory**, Interscience, 1963.

Coulson, C.A., *Momentum distribution in molecular systems. I. Single bond. III.*

Bonds of higher order, Proc. Camb. Phys. Soc., **37** 74, 1941.

Coulson, C.A. and Duncanson, W.E., *Momentum distribution in molecular systems. II. C and C-H bond*, Proc. Camb. Phys. Soc., **37** 67, 1941.

Coxeter, H.S.M. et al. Editors, **Mathematical Expositions, Volume 21: The Collected Papers of Alfred Young, 1873-1940**, University of Toronto Press, Toronto, 1977.

Dahl, J.P., *The Wigner function*, Physica A, **114** 439, 1982.

Dahl, J.P., *On the group of translations and inversions of phase space and the Wigner function*, Phys. Scripta, **25** 499-503, 1982.

Dahl, J.P., *Dynamical equations for the Wigner functions*, in **Energy Storage and Redistribution in Molecules**, Hinze, J., Ed., Plenum, New York, 1983.

Dahl, J.P., *The phase-space representation of quantum mechanics and the Bohr-Heisenberg correspondence principle*, in **Semiclassical Description of Atomic and Nuclear Collisions**, Bang, J. and De Boer, J., Eds., North Holland, Amsterdam, 1985.

Dahl, J.P., *The dual nature of phase-space representations*, in **Classical and Quantum Systems**, Doebner, H.D. and Schroeck, F., Jr., Eds., World Scientific, Singapore, 1993.

Dahl, J.P., *A phase space essay*, in **Conceptual Trends in Quantum Chemistry**, Kryachko, E.S. and Calais, J.L., Eds., Kluwer Academic Publishers, Dordrecht, Netherlands, 1994.

Dahl, J.P. and Springborg, M., *The Morse oscillator in position space, momentum space, and phase space* J. Chem. Phys., **88** 4535-47, 1988.

Dahl, J.P. and Springborg, M., *Wigner's phase-space function and atomic structure. I. The hydrogen atom* , J. Mol. Phys., **47** 1001, 1982.

Das, G.P.; Ghosh, S.K.; and Sahni, V.C., *On the correlation energy density functional in momentum space*, Solid State Commun., **65** (7) 719-21, 1988.

Dinesha, K.V., Rettrup, S., and Sarma, C.R., *An Indexing Scheme for Spin-Free Configurations of Electrons*, Int. J. Quantum Chem., **34** 445-455, 1988.

Dinesha, K.V., Sarma, C.R, and Rettrup, S., *Group Theoretical Techniques and the Many-Electron Problem*, Adv. Quantum Chem., **14** 125-168, P. 1981.

Englefield, M.J., **Group theory and the Coulomb problem**, Wiley-Interscience, New York, 1972.

Fock, V.A., *Zur Theorie des Wasserstoffatoms*, Z. Phys., **98** 145, 1935.

Fock, V.A., *Hydrogen atoms and non-Euclidian geometry*, Kgl. Norske Videnskab. Forh., **31** (1958) 138.

Fonseca, A.C. and Pena, M.T., *Rotational-invariant Sturmian-Faddeev ansatz for the solution of hydrogen molecular ion (H2+): a general approach to molecular three- body problems*, Phys. Rev. A: Gen. Phys., **38** 4967-84, 1988.

Friis-Jensen, B., Cooper, D.L., and Rettrup, S., *Normalization of Projected Spin Eigenfunctions*, J. Math. Chem., **22** 249-254, 1997.

Friis-Jensen, B., Cooper, D.L., and Rettrup, S., *Spin-Coupled Calculations based on Projected Spin Eigenfunctions*, Theor. Chem. Acc., **99** 64-67, 1998.

Friis-Jensen, B., and Rettrup, S., *A Spin-Free Approach for Evaluation of Elec-*

tronic Matrix Elements using Character Operators of S_N, Int. J. Quantum Chem., **60**, 983-991, 1996.

Friis-Jensen, B., Rettrup, S., and Sarma, C.R. *An Indexing Scheme for Classes of S_N: Partitions of N*, Int. J. Quantum Chem., **64** 421-426, 1997.

Gasaneo, G. et al., *Theory of hyperspherical Sturmians for three-body reactions*, J. Phys. Chem. A, **113**, 14573-14582, 2009.

Gasaneo, G. et al., *Multivariable hypergeometric solutions for three charged particles*, J. Phys. B, **30** L265-L271, 1997.

Gasaneo, G., and Mmacek, J.H., *Hyperspherical adiabatic eigenfunctions for zero-range potentials*, J. Phys. B., **35** 2239-2250, 2002.

Gasaneo, G., and Colavecchia, F.D., *Two-body Coulomb wave functions as a kernel for three alternative integral transformations*, J. Phys. A, **36** 8443-8462, 2003.

Gerratt, J., *General theory of spin-coupled wave functions for atoms and molecules*, Adv. At. Mol. Phys., **7** 141-221, 1971.

Goscinski, O., *Conjugate eigenvalue problems and the theory of upper and lower bounds*, Adv. Quantum Chem., **41** 51-85, 2003.

Gradshteyn, I.S. and Ryshik, I.M., **Tables of Integrals, Series and Products**, Academic Press, New York, 1965.

Griffeth, J.S., **The Theory of Transition Metal Ions**, Cambridge, 1964.

Gruber, B., Editor, **Symmetries in Science, VI: From the Rotation Group to Quantum Algebras**, Plenum, New York, 1993.

Guldberg, A., Rettrup, S., Bendazzoli, G.L., and Palmieri, P., *A New Symmetric Group Programme for Direct Configuration Interaction Studies in Molecules*, Int. J. Quantum Chem. (Symp.), **21** 513-521, 1987.

Hall, G.G., **Applied Group Theory**, Longmans, 1967.

Hammermesh, M., **Group Theory and Its Applications to Physical Problems**, Addison-Wesley, 1964.

Hansen, T.B., **The many-center one-electron problem in momentum space**, Thesis, Chemical Institute, University of Copenhagen, 1998.

Harris, F.E. and Michels, H.H., *Evaluation of molecular integrals for Slater-type orbitals*, Adv. Chem. Phys., **13** 205, 1967.

Hartt, K. and Yidana, P.V.A., *Analytic Sturmian functions and convergence of separable expansions*, Phys. Rev. C: Nucl. Phys., **36** 475-84, 1987.

Heine, V., **Group Theory in Quantum Mechanics**, Pergamon, 1960.

Herschbach, D. R., *Dimensional interpolation for two-electron atoms*, J. Chem. Phys., **84** 838, 1986.

Herschbach, D. R., Avery, J. and Goscinski, O., Eds., **Dimensional Scaling in Chemical Physics**, Kluwer, Dordrecht, 1993.

Hochstrasser, R.M., **Molecular Aspects of Symmetry**, Benjamin, 1966.

Hollas, J.M., **Symmetry in Molecules**, Chapman, 1972.

Jaffé, H.H. and Ochrin M., **Symmetry in Chemistry**, Wily, 1965.

Judd, B.R., **Angular Momentum Theory for Diatomic Molecules**, Academic Press, New York, 1975.

Judd, B.R., **Operator Techniques in Atomic Spectroscopy**, McGraw-Hill, 1963.

Judd, B.R., **Second Quantization and Atomic Spectroscopy**, Johns Hopkins Press, 1967.

Kadolkar, C., Sarma, C.R., and Rettrup, S., *Configuration Interaction Studies Using Biorthogonal Approach to VB Basis*, Int. J. Quantum Chem., **53** 183-187, 1995.

Kaplan, I.G., **Symmetry of Many-Electron Systems**, Academic Press, 1975.

Karule, E. and Pratt, R.H., *Transformed Coulomb Green function Sturmian expansion*, J. Phys. B: At., Mol. Opt. Phys., **24** 1585-91, 1991.

Knox, R.S. and Gold, A., **Symmetry in the Solid State**, Benjamon, 1964.

Koga, T. and Murai, Takeshi, *Energy-density relations in momentum space. III. Variational aspect*, Theor. Chim. Acta, **65** 311-16, 1984.

Koga, T., *Direct solution of the $H(1s) - H^+$ long-range interaction problem in momentum space*, J. Chem. Phys., **82** 2022, 1985.

Koga, T. and Matsumoto, S., *An exact solution of the interaction problem between two ground-state hydrogen atoms*, J. Chem. Phys., **82** 5127, 1985.

Koga, T. and Kawaai, R., *One-electron diatomics in momentum space. II. Second and third iterated LCAO solutions* J. Chem. Phys., **84** 5651-4, 1986.

Koga, T. and Matsuhashi, T., *One-electron diatomics in momentum space. III. Nonvariational method for single-center expansion*, J. Chem. Phys., **87** 1677-80, 1987.

Koga, T. and Matsuhashi, T., *Sum rules for nuclear attraction integrals over hydrogenic orbitals*, J. Chem. Phys., **87** 4696-9, 1987.

Koga, T. and Matsuhashi, T., *One-electron diatomics in momentum space. V. Nonvariational LCAO approach*, J. Chem. Phys., **89** 983, 1988.

Koga, T.; Yamamoto, Y., and Matsuhashi, T., *One-electron diatomics in momentum space. IV. Floating single-center expansion*, J. Chem. Phys., **88** 6675-6, 1988.

Koga, T. and Ougihara, T., *One-electron diatomics in momentum space. VI. Nonvariational approach to excited states*, J. Chem. Phys., **91** 1092-5, 1989.

Koga, T.; Horiguchi, T. and Ishikawa, Y., *One-electron diatomics in momentum space. VII. Nonvariational approach to ground and excited states of heteronuclear systems*, J. Chem. Phys., **95** 1086-9, 1991.

Kupperman, A. and Hypes, P.G., *3-dimensional quantum mechanical reactive scattering using symmetrized hyperspherical coordinates*, J. Chem. Phys., **84** 5962, 1986.

la Cour Jansen, T., Rettrup, S., Sarma, C.R., and Snijders, J.G. *On the Evaluation of Spin-Orbit Coupling Matrix Elements in a Spin Adapted Basis*, Int. J. Quantum Chem., **73** 23-27, 1999.

Leech, J.W. and Newman, D.J., **How to Use Groups**, Methuen, 1969.

Lin, C.D., *Analytical channel functions for 2-electron atoms in hyperspherical coordinates*, Phys. Rev. A, **23** 1585, 1981.

Linderberg, J. and Öhrn, Y., *Kinetic energy functional in hyperspherical coordinates*, Int. J. Quantum Chem., **27** 273, 1985.

Louck, J.D., *Generalized orbital angular momentum and the n-fold degenerate quantum mechanical oscillator*, J. Mol. Spectr., **4** 298, 1960.

Loebl, E.M., Editor, **Group Theory and Its Applications**, Academic Press,

1968.

Lomont, J.S., **Applications of Finite Group Theory**, Academic Press, 1959.

Manakov, N.L.; Rapoport, L.P. and Zapryagaev, S.A., *Sturmian expansions of the relativistic Coulomb Green function*, Phys. Lett. A, **43** 139-40, 1973.

Maquet, Alfred; M., Philippe and Veniard, Valerie, *On the Coulomb Sturmian basis*, NATO ASI Ser, Ser C, **271** (Numer. Determ. Electron. Struct. At., Diat. Polyat. Mol.), 295-9, 1989.

McWeeny, R., and Coulson, C.A., *The computation of wave functions in momentum space. I. The helium atom*, Proc. Phys. Soc. (London) A, **62** 509, 1949.

McWeeny, R., *The computation of wave functions in momentum space. II. The hydrogen molecule ion*, Proc. Phys. Soc. (London) A, **62** 509, 1949.

McWeeny, R., **Methods of Molecular Quantum Mechanics**, (second edition), Academic Press, 1992.

McWeeny, R., **Symmetry**, Pergamon, 1963.

Meijer, P.H.E. and Bauer, E., **Group Theory**, North Holland, 1962.

Michels, M.A.J., *Evaluation of angular integrals in rotational invariants*, Int. J. Quantum Chem., **20** 951-952, 1981.

Monkhorst, H.J. and Harris, F.E., *Accurate calculation of Fourier transform of two-center Slater orbital products*, Int. J. Quantum Chem., **6** 601, 1972.

Monkhorst, H.J. and Jeziorski, B., *No linear dependence or many-center integral problems in momentum space quantum chemistry*, J. Chem. Phys., **71** 5268-9, 1979.

Moore, C.E., **Atomic Energy Levels; Circular of the National Bureau of Standards 467**, Superintendent of Documents, U.S. Government Printing Office, Washington 25 D.C., 1949.

Morse P.M. and Feshbach, H., **Methods of Theoretical Physics**, McGraw Hill, 1953.

Motenberg, M., Bivens, R., Metropolis, N., and Wooten, J.K. Jr. **The 3-j and 6-j Symbols**, Technology Press, MIT, Cambridge Mass., 1959.

Murnagham, F.D. **The Theory of Group Representations**, Johns Hopkins Press, Baltimore, 1938.

National Institute for Fusion Science (NIFS), *Atomic and Molecular Databases: Data for Autoionizing States*, http://dprose.nifs.ac.jp/DB/Auto/

National Institute of Standards and Technology (NIST), *NIST Atomic Spectra Database*, http://physics.nist.gov/asd .

National Institute of Standards and Technology (NIST), *CODATA Recommended Values of the Fundamental Physical Constants: 2002*, http://physics.nist.gov/cuu/Constants/

Navasa, J. and Tsoucaris, G., *Molecular wave functions in momentum space*, Phys. Rev. A, **24** 683, 1981.

Ojha, P.C., *The Jacobi-matrix method in parabolic coordinates: expansion of Coulomb functions in parabolic Sturmians*, J. Math. Phys. (N Y), **28** 392-6, 1987.

Orchin, M. and Jaffé, H.H., **Symmetry, Orbitals and Spectra**, Wily, 1971.

Orgel, L.E., **An Introduction to Transition Metal Chemistry Ligand**

Field Theory, Methuen, 1960.

Paldus, J., *Coupled Cluster Methods*, in Wilson, S., Editor, **Handbook of Molecular Physics and Quantum Chemistry**, Wiley, 2003.

Paldus, J., *Dynamical Groups*, in Drake, G.W.F. Editor, **Atomic, Molecular and Optical Physics Handbook**, American Institute of Physics, New York, 1996.

Paldus, J., and Gould, M.D., *Unitary Group Approach to Reduced Density Matrices, II*, Theor. Chim. Acta, **86** 83-96, 1993.

Paldus, J., Rettrup, S., and Sarma, C.R., *Clifford algebra realization of Rumer-Weyl basis*, J. Mol. Structure (Theochem), **199** 85-101, 1989.

Palmieri, P., and Rettrup, S., *Teoria Del Gruppo Simmetrico Per Il Calcolo Degli Elementi Di Matrice E Per La Soluzione Delle Equazioni CI*, in **Scuola di Chimica Teorica : Basi Teoriche ed Applicationi Numeriche**, Moccia, R. and Pisani, C., Eds., Polo Editorale Chimico, Milano, 197-276, 1990.

Palmieri, P., Tarroni, R., Mitrushenkov, A.O., and Rettrup, S., *Efficient Truncation Strategies for Multi Reference Configuration Interaction Molecular Energies and Properties*, J. Chem. Phys., **109** 7085-7091, 1998.

Pauncz, R., **The Symmetric Group in Quantum Chemistry**, CRC Press, Boca Raton, 1995.

Petrochen, M.I. and Trifanov, E.D., **Group Theory in Quantum Mechanics**, Iliffe, 1969.

Pinchon, D. and Hoggan, P.E., *Rotation matrices for real spherical harmonics: general rotations of atomic orbitals in space-fixed axes*, J. Phys. A., **40** 1597-16100, 2007.

Pinchon, D. and Hoggan, P.E., *Gaussian approximation of exponential type orbitals based on B functions*, Int. J. Quantum Chem., **109** 135-148, 2009.

Rettrup, S., *A recursive formula for Young's orthogonal representation*, Chem. Phys. Lett., **47** 59-60, 1977.

Rettrup, S., *Many-body correlations using unitary groups*, in **The Unitary Group for the Evaluation of Electronic Energy Matrix Elements**, Lecture Notes in Chemistry Vol. **22**, 108-118, Hinze, J., Ed., Springer-Verlag, Berlin, 1981.

Rettrup, S., *An Iterative Method for Calculating Several of the Extreme Eigensolutions of Large Real Non-Symmetric Matrices*, J. Comp. Phys., **45** 100-107, 1982.

Rettrup, S., *Direct evaluation of spin representation matrices and ordering of permutation group elements*, Int. J. Quantum Chem., **29** 119-128, 1986.

Rettrup, S., *The permutation group in many-electron theory*, in **Conceptual Trends in Quantum Chemistry III**, Kryachko, E.S., and Calais, J.L., Eds, Kluwer, Dordrecht, 225-238, 1997.

Rettrup, S., **Lecture Notes on Quantum Chemistry**, Department of Chemistry, University of Copenhagen, 2003.

Rettrup, S., *Alternative graphical representation of determinential many-electron states*, Int. J. Quantum Chem., **106** 2511-2517, 2006.

Rettrup, S., Bendazzoli, G.L., Evangelisti, S., and Palmieri, P., *A symmet-

ric group approach to the calculation of electronic correlation effects in molecules, in **Understanding Molecular Properties**, 533-546, Avery, J.S., Dahl, J.P., and Hansen, Aa.E., Eds., Reidel Publ. Co., Dordrecht, Holland, 1987.

Rettrup, S., and Chisholm, C.D.H., *On R(4) Symmetries in atomic structure*, Theoret. Chim. Acta (Berl.), **57** 209-218, 1980.

Rettrup, S., and Pauncz, R., *Representations of the symmetric group generated by projected spin functions: a graphical approach*, Int. J. Quantum Chem., **60** 91-98, 1996.

Rettrup, S., and Sarma, C.R., *A new program for CI calculations in molecules*, Theoret. Chim. Acta (Berl.), **46** 73-76, 1977.

Rettrup, S., and Sarma, C.R., *A Programmable procedure for generating many-particle states*, Phys. Letters, **75A** 181-182, 1980.

Rettrup, S., Sarma, C.R., and Dahl, J.P., *Molecular point group adaptation of spin-free configurations*, Int. J. Quantum Chem., **22** 127-148, 1982.

Rettrup, S., Thorsteinsson, T., and Sarma, C.R., *A graphical approach to configuration interaction studies in molecules using determinants of non-orthogonal orbitals*, Int. J. Quantum Chem., **40** 709-717, 1991.

Robinson, G. de B., **Representation Theory of Symmetric Groups**, University of Toronto Press, Toronto, 1961.

Rohwedder, B. and Englert, B.G., *Semiclassical quantization in momentum space*, Phys. Rev. A: At., Mol., Opt. Phys., **49** 2340-6, 1994.

Rose, M.E., **Elementary Theory of Angular Momentum**, Wiley, 1957.

Rotenberg, M., Ann. Phys. (New York), **19** 262, 1962.

Rotenberg, M., *Theory and application of Sturmian functions*, Adv. At. Mol. Phys., **6** 233-68, 1970.

Royer, A., *Wigner function as the expectation value of a parity operator*, Phys. Rev. A, **15** 449-450, 1977.

Rudin, W., **Fourier Analysis on Groups**, Interscience, New York, 1962.

Rutherford, D.E., **Substitutional Analysis**, Edinburgh University Press, Edinburgh, 1948.

Sarma, C.R., Ahsan, M.A.H., and Rettrup, S., *A graphical approach to permutation group representations for many-electron systems*, Int. J. Quantum Chem., **58** 637-643, 1996.

Sarma, C.R., Palmieri, P., and Rettrup, S., *A graphical scheme for addressing determinants in large scale configuration interaction treatments*, Mol. Phys., **98** 1851-1856, 2000.

Sarma, C.R., Ravi Shankar, N., and Rettrup, S., *A graphical scheme for representing many-electron spin-free configurations*, Int. J. Quantum Chem., **86** 417-421, 2002.

Sarma, C.R., and Rettrup, S., *A programmable spin-free method for configuration interaction*, Theoret. Chim. Acta (Berl.), **46** 63-72, 1977.

Sarma, C.R., and Rettrup, S., *A programme for the study of many-body correlations: matrix elements of the generators of U(n)*, J. Phys. A: Math. Gen., **13** 2267-2274, 1980.

Schönland D.S., **Molecular Symmetry**, Van Norstrand, 1965.

Shibuya, T. and Wulfman, C.E., Molecular orbitals in momentum space, *Proc. Roy. Soc. A*, **286** (1965) 376.

Shull, H. and Löwdin, P.-O., *Superposition of configurations and natural spin-orbitals. Applications to the He problem*, J. Chem. Phys., **30** 617, 1959

Sinanoglu, O., and Bruekner, K., **Three Approaches to Electron Correlation in Atoms**, Yale University Press, 1970.

Slater, J.C., **Quantum Theory of Atomic Structure**, McGraw-Hill, 1960.

Smith, F.T., *Generalized angular momentum in many-body collisions*, Phys. Rev., **120** 1058, 1960.

Smith, F.T., *A symmetric representation for three-body problems. I. Motion in a plane*, J. Math. Phys., **3** 735, 1962.

Smith, F.T., *Participation of vibration in exchange reactions*, J. Chem. Phys., **31** 1352-9, 1959.

Smith, V.H. Jr., *Density functional theory and local potential approximations from momentum space considerations*, in **Local Density Approximations in Quantum Chemistry and Solid State Physics**, Plenum, New York, N. Y, Dahl, J.P. and Avery, J.,Eds., 1-19, 82, 1984.

Springborg, M. and Dahl, J.P., *Wigner's phase-space function and atomic structure*, Phys. Rev. A, **36** 1050-62, 1987.

Sommer-Larsen. P., Ph.D. Thesis, University of Copenhagen, 1990.

Sugano, S., Tanabe, Y. and Kamimura, H., **Multiplets of Transition Metal Ions in Crystals**, Academic Press, 1960.

Suzuki, H., **Electronic Absorption Spectra and Geometry of Organic Molecules**, Academic Press, 1967.

Thorsteinsson, T., and Rettrup, S., *Configuration interaction studies of electronic structure*, in **Supercomputing Annual Report 1997**, J. Wasniewski, Ed., UNI•C, The Danish Computing Centre for Research and Education, Lyngby, 1998.

Tinkham, M., **Group Theory and Quantum Mechanics**, McGraw-Hill, 1964.

Urch, D.S., **Orbitals and Symmetry**, Penguin, 1970.

Vilenkin, N.K., **Special Functions and the Theory of Group Representations**, American Mathematical Society, Providence, R.I., 1968.

Watanabe, H., **Operator Methods in Ligand Field Theory**, Prentace Hall, 1966.

Weatherford, C.A., and Jones, H.W., Editors, **International Conference on ETO Multicenter Integrals**, Reidel, Dordrecht, 1982.

Weyl, H., **The Theory of Groups in Quantum Mechanics**, Dover, 1950.

Weyl, H., **The Classical Groups, Their Invariants and Representations**, Princeton University Press, 1939.

Wen, Z.-Y. and Avery, J., *Some properties of hyperspherical harmonics*, J. Math. Phys., **26** 396, 1985.

Weniger, E.J., *Weakly convergent expansions of a plane wave and their use in Fourier integrals*, J. Math. Phys., **26** 276, 1985.

Weniger, E.J. and Steinborn, E.O., *The Fourier transforms of some exponential-type basis functions and their relevance for multicenter problems*, J. Chem. Phys., **78** 6121, 1983.

Weniger, E.J.; Grotendorst, J., and Steinborn, E.O., *Unified analytical treatment of overlap, two-center nuclear attraction, and Coulomb integrals of B functions via the Fourier transform method*, Phys. Rev. A, **33** 3688, 1986.

Wigner, E.P., **Group Theory and Its Application to the Quantum Mechanics of Atomic Spectra**, Academic Press, 1959.

Wilson, S., Editor, **Handbook of Molecular Physics and Quantum Chemistry**, Wiley, 2003.

Wulfman, C.E., *Semiquantitative united-atom treatment and the shape of triatomic molecules*, J. Chem. Phys., **31** 381, 1959.

Wulfman, C.E., *Dynamical groups in atomic and molecular physics*, in **Group Theory and its Applications**, Loebel, E.M. Ed., Academic Press, 1971.

Wulfman, C.E., *Approximate dynamical symmetry of two-electron atoms*, Chem. Phys. Lett., **23** 370-372, 1973.

Wulfman, C.E., *On the space of eigenvectors in molecular quantum mechanics*, Int. J. Quantum Chem., **49** 185, 1994.

Wulfman, C.E., **Dynamical Symmetry**, World Scientific, 2011.

Yurtsever, E.; Yilmaz, O., and Shillady, D.D., *Sturmian basis matrix solution of vibrational potentials*. Chem. Phys. Lett., **85** 111-116, 1982.

Yurtsever, E., *Franck-Condon integrals over a Sturmian basis. An application to photoelectron spectra of molecular hydrogen and molecular nitrogen*, Chem. Phys. Lett., **91** 21-26, 1982.

Index